高等学校计算机基础教育规划教材

Java Web 应用开发基础教程

郭庆 田甜 王向辉 徐遵义 编著

清华大学出版社
北京

内容简介

本书分为13章：第1章主要介绍了基于JDK 1.7、Tomcat 7、MySQL 5、Eclipse JEE搭建Java Web集成开发环境、Eclipse集成Tomcat、Eclipse的基本使用；第2～10章介绍了Java Web应用开发中的常用官方技术标准，包括JDBC、Servlet、JSP、JavaBean、表达式语言、JSTL、过滤器和监听器、自定义标签、JPA；第11～13章介绍了目前软件开发企业使用较多的开源框架Spring，包括Spring框架核心基础、使用Spring持久化、Spring MVC。

本书在详细说明Java Web应用开发相关技术的理论和API基础上，结合第1章搭建的集成开发环境编写了大量工程性和系统性案例。掌握这些案例有助于深入理解Java Web相关技术的实际应用场景。

本书可作为高等学校Java Web应用开发课程的教材，也可供从事软件开发的工程技术人员自学使用。

图书在版编目(CIP)数据

Java Web应用开发基础教程/郭庆等编著. —北京：清华大学出版社，2018(2024.7重印)
(高等学校计算机基础教育规划教材)
ISBN 978-7-302-50736-9

Ⅰ. ①J… Ⅱ. ①郭… Ⅲ. ①JAVA语言—程序设计—高等学校—教材 Ⅳ. ①TP312.8

中国版本图书馆CIP数据核字(2018)第172726号

责任编辑：袁勤勇　李　晔
封面设计：常雪影
责任校对：时翠兰
责任印制：宋　林

出版发行：清华大学出版社
　网　　址：https://www.tup.com.cn，https://www.wqxuetang.com
　地　　址：北京清华大学学研大厦A座　　**邮　　编**：100084
　社 总 机：010-83470000　　**邮　　购**：010-62786544
　投稿与读者服务：010-62776969，c-service@tup.tsinghua.edu.cn
　质量反馈：010-62772015，zhiliang@tup.tsinghua.edu.cn
　课件下载：https://www.tup.com.cn，010-83470236
印 装 者：三河市龙大印装有限公司
经　　销：全国新华书店
开　　本：185mm×260mm　　**印　张**：20.5　　**字　　数**：473千字
版　　次：2018年10月第1版　　**印　　次**：2024年7月第8次印刷
定　　价：59.00元

产品编号：075055-02

前言

在服务器端编程方面，Java Web 技术以其可靠性高、安全性好等特点成为市场主流，因此深入学习和掌握 Java Web 应用开发的相关技术变得日益重要。

笔者从事 Java、Java EE、Java Web 等相关课程的教学和培训 10 余年，主持和参与了多项基于 Java 技术的项目，积累了许多教学和实践经验。本书致力于为学习 Java Web 应用开发技术的读者提供一个坚实的理论基础和实践基础，在注重基础知识理解的同时强调实践能力的培养。本书具有如下特点：

(1) 内容简洁、完整，便于读者全面掌握 Java Web 应用开发技术的基础知识。

(2) 实践性强，以实例形式介绍多种 Java Web 应用开发技术。

(3) 系统性强，以实际常用的集成开发环境 Eclipse、Web 服务器 Tomcat、数据库 MySQL 等软件为主进行了整合，并针对 Java Web 应用开发中的常用技术编写了大量系统性案例。

(4) 侧重于服务器后端技术，详细介绍 Java Web 应用开发相关技术。

(5) 思考与习题基础性强，有助于提高读者兴趣，解决新手编程入门难的问题。

全书共分 13 章，内容包括：集成开发环境搭建与使用、JDBC、Servlet、JSP、JavaBean、表达式语言、JSTL、过滤器和监听器、自定义标签、JPA、Spring 框架核心基础、使用 Spring 持久化、Spring MVC。

本书既适合作为高等院校 Java Web 应用开发技术课程的教材，也可作为高职高专、成人高等教育、社会培训班的教材，还可作为 Java Web 应用开发技术自学者的教材或参考书。

本书由山东建筑大学的郭庆、田甜、王向辉、徐遵义编著。其中，徐遵义编写了第 1、9 章，郭庆编写了第 2～4 章和第 10～13 章，田甜编写了第 5、6 章，王向辉编写了第 7、8 章。

本书参考了大量的国内外文献，在此向这些文献的作者表示深深的敬意和衷心的感谢！

软件技术发展迅速，加之作者水平和时间有限，书中难免有不妥之处，恳请读者批评指正。

编　者

2018 年 1 月

目录

第1章

集成开发环境搭建

现在 Java Web 应用开发中常见的集成开发软件主要有开源软件 Eclipse jee(以下简称 Eclipse)和 NetBeans、商业软件 MyEclipse 和 IntelliJ 等。其中,Eclipse 使用者众多。

Java Web 应用开发集成环境搭建通常包括安装 JDK、Tomcat、Eclipse、数据库服务器 MySQL,并需要对安装的某些软件进行配置。另外还需要在 Eclipse 中集成 Tomcat。

本书的 Java Web 集成开发环境基于 Windows 10 64 位操作系统,因此 JDK、Eclipse、Tomcat 均采用的是 64 位版本。对于 32 位 Windows,请到相关网站下载对应的 32 位版本的软件。

1.1 集成开发环境搭建

1.1.1 安装 JDK

1. 安装 JDK

双击 jdk-7u75-windows-x64.exe 文件进行安装。默认安装路径 C:\Program Files\Java\ jdk1.7.0_75,也可以在安装时修改为其他路径。

2. 配置 JDK

(1) 新增系统变量 JAVA_HOME。在 Windows 10 下,右击"此电脑"→"属性"→"高级系统设置",弹出"系统属性"对话框,如图 1-1 所示。

在图 1-1 中,选择"高级"→"环境变量",在"系统变量"中单击"新建"按钮,弹出"新建系统变量"对话框,如图 1-2 所示。

在图 1-2 中,在"变量名"文本框中输入 JAVA_HOME,在"变量值"文本框中输入 JDK 安装路径,例如 C:\Program Files\Java\jdk1.7.0_75。

依次单击"新建系统变量"对话框、"环境变量"对话框、"系统属性"对话框中的"确定"按钮,完成新增系统变量 JAVA_HOME 的操作。

(2) 编辑系统变量 Path。在 Windows 10 下,单击"此电脑"→"属性"→"高级系统设

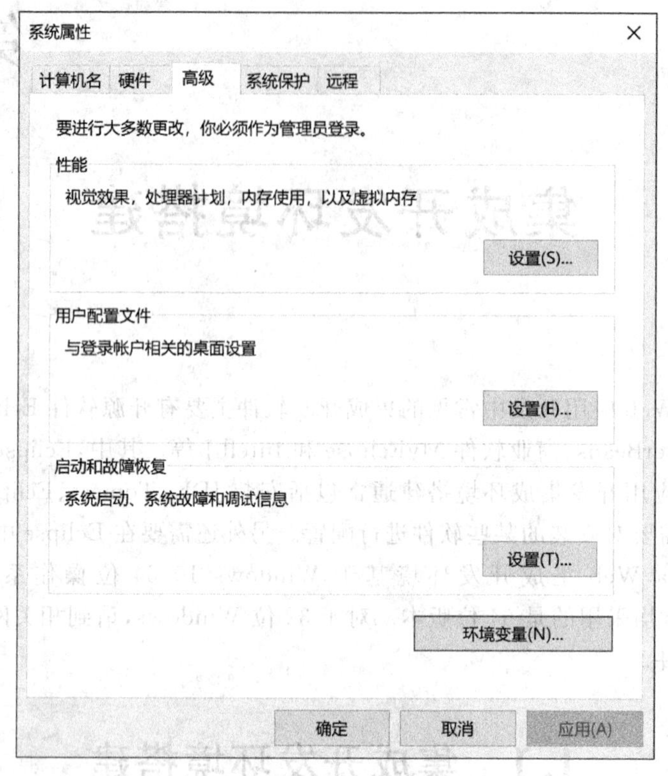

图 1-1 “系统属性”对话框

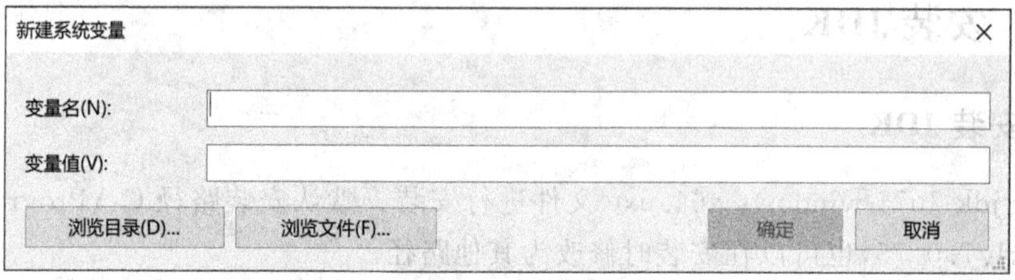

新建系统变量
变量名(N):
变量值(V):
浏览目录(D)...　浏览文件(F)...　确定　取消

图 1-2 “新建系统变量”对话框

置”，选择“高级”→“环境变量”；选择“系统变量”下方窗口中的 Path 变量后，单击“编辑”按钮打开“编辑系统变量”对话框，单击“新建”按钮并输入%JAVA_HOME%\bin。

依次单击“编辑环境变量”对话框、“环境变量”对话框、“系统属性”对话框中的“确定”按钮，完成编辑系统变量 Path 的操作。

3. 测试 JDK 安装和配置是否成功

在 Windows 10 桌面左下方的搜索框输入 cmd，选择“命令提示符”。在“命令提示符”窗口输入命令 javac 后并按 Enter 键，如果显示如图 1-3 所示的 javac 使用帮助，说明 JDK 安装和配置成功了。

```
C:\Users\guoqing>javac
用法: javac <options> <source files>
其中, 可能的选项包括:
  -g                          生成所有调试信息
  -g:none                     不生成任何调试信息
  -g:{lines,vars,source}      只生成某些调试信息
  -nowarn                     不生成任何警告
```

图 1-3　在“命令提示符”窗口输入命令 javac 后的显示效果

1.1.2　安装 Tomcat

Tomcat 官方网站提供了两种类型的文件供用户下载。.exe 类型的文件是 Windows 下的可执行安装文件;.zip 类型的文件属于绿色软件,不用安装,解压即可使用。本书使用的是.zip 类型的文件。

1. 安装 Tomcat

将压缩文件 apache-tomcat-7.0.73.zip 解压到某个路径,例如 C:\JavaProgram\Tomcat\apache-tomcat-7.0.73\,即 Tomcat 的安装主目录为 C:\JavaProgram\Tomcat\的 apache-tomcat-7.0.73。在 Windows 10 资源管理器中,apache-tomcat-7.0.73 目录显示效果如图 1-4 所示。

名称	日期
bin	17/11/13 09:19
conf	17/11/13 09:19
lib	17/11/13 09:19
logs	18/3/6 11:55
temp	17/11/13 09:19
webapps	18/3/6 11:58
work	17/11/13 09:19
apache-tomcat-7.0.73-src	16/12/23 14:41
LICENSE	16/11/7 23:29
NOTICE	16/11/7 23:29
RELEASE-NOTES	16/11/7 23:29
RUNNING	16/11/7 23:29

图 1-4　apache-tomcat-7.0.73 目录显示效果

2. 测试 Tomcat 安装是否成功

首先到 apache-tomcat-7.0.73 的 bin 目录,双击 startup.bat,即可启动 Tomcat 服务器。启动 Tomcat 后,任务栏会显示 Tomcat DOS 窗口;请不要关闭该窗口。关闭该窗口将导致关闭 Tomcat。接着打开一个浏览器窗口,并在地址栏输入 http://localhost:8080 后按 Enter 键,如果浏览器显示 Tomcat 欢迎画面,如图 1-5 所示,说明 JDK、Tomcat 均已安装和配置成功。如果要关闭 Tomcat,请双击 bin 目录下的 shutdown.bat。

图 1-5　Tomcat 欢迎画面

1.1.3　安装 MySQL

解压文件 mysql-5.5.18-win32.rar 后，双击 MySQL 的安装文件 mysql-5.5.18-win32.msi 进行安装。默认安装路径为 C:\Program Files (x86)\MySQL\MySQL Server 5.5。为了与本书一致，注意在安装过程中将 MySQL 的登录账号 root 的密码设置为 root，字符编码改为 utf8(即 UTF-8)。

1.1.4　安装 Eclipse

1. 安装文件

解压文件 eclipse-jee-mars-2-win32-x86_64.zip 到 C:\JavaProgram\eclipse-jee-mars-2-win32-x86_64\即可完成 Eclipse 的安装。

2. 配置工作空间

Eclipse 安装结束后，在 C:\JavaProgram\eclipse-jee-mars-2-win32-x86_64\下创建文件夹 workspaces 作为工作空间。工作空间用于管理项目。在 Eclipse 平台中，要编写 Java 类需要首先创建项目。项目是 Eclipse 平台的基本工作单元。工作空间可以位于其他路径，如 F:\workspaces 或者 E:\MyApps。

第一次启动 Eclipse，会弹出设置工作空间的窗口。将工作空间设置为刚刚创建的 workspaces 的路径 C:\JavaProgram\eclipse-jee-mars-2-win32-x86_64\workspaces。

3. 设置 Eclipse 的默认 JRE

通常需要将 Eclipse 的默认 JRE 设置为前面安装的 JDK。

在 Eclipse 中，选择 Window→Preference，单击左侧的 Java→Installed JREs 节点，单击右方窗口中的 Add 按钮，在弹出的 JRE Type 窗口选择 Standard VM，单击 Next 按钮后弹出 JRE Definition 窗口，如图 1-6 所示。

在 JRE Definition 窗口中，单击 JRE home 右边 Directory 按钮浏览 JDK 的主目录 C:\Program Files\Java\jdk1.7.0_75，单击 OK 按钮后，在 JRE Definition 窗口单击 Finish 按钮，最后单击 Installed JREs 窗口下方的 OK 按钮，完成 Eclipse 默认 JRE 的设置。

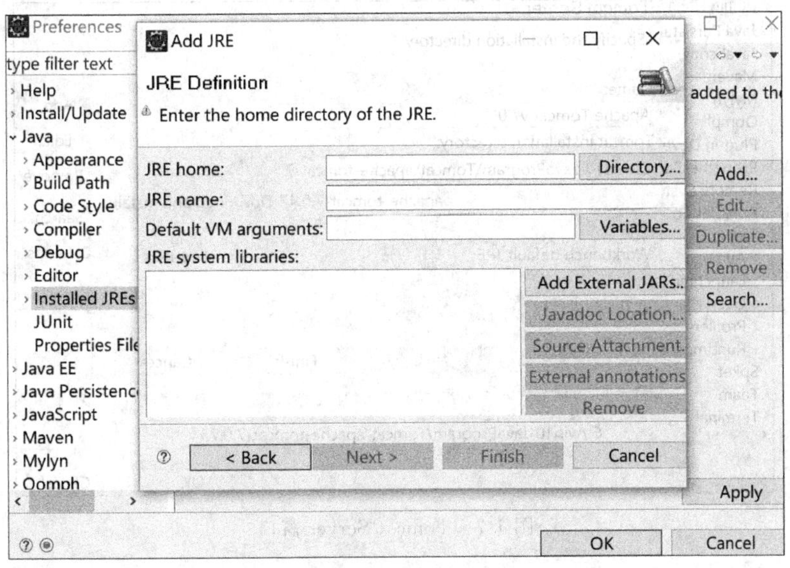

图 1-6　JRE Definition 窗口

1.1.5　Eclipse 集成 Tomcat

前面已经成功安装了 Tomcat 和 Eclipse。实际开发中如果总是到%Tomcat%\bin\双击 startup. bat 和 shutdown. bat 来启动和关闭 Tomcat，显得太麻烦且不能同步调试代码。Eclipse 提供了集成 Tomcat 的插件，可以实现在 Eclipse 中启动和关闭 Tomcat，进行 Web 应用程序的调试。

1. Eclipse 集成 Tomcat

Eclipse 集成 Tomcat 需要进行以下两步操作。

(1) 配置服务器运行环境。依次选择 Window→Preference→Server→Runtime Environment，在 Server Runtime Environments 右上方窗口单击 Add 按钮；在 New Server Runtime Environment 窗口选择 Apache Tomcat 7. 0，单击 Next 按钮，弹出 Tomcat Server 窗口，如图 1-7 所示。

在 Tomcat Server 窗口中，单击 Browse 按钮浏览 Tomcat 7. 0. 73 的安装主目录 C:\Win10-JavaProgram\Tomcat\apache-tomcat-7. 0. 73 并选中，单击下方的 Finish 按钮。

在 Server Runtime Environments 窗口中单击 OK 按钮完成服务器运行环境 Tomcat 7. 0. 73 配置。

(2) 新建服务器。在 Eclipse 下方的 Servers 窗口中，如图 1-8 所示，右击并选择 New →Server 命令，打开 New Server 向导。

在 New Server 向导打开的 Define a New Server 窗口(如图 1-9 所示)中，选择 Tomcat v7. 0 Server，其 Server runtime environment 为前面配置的 Apache Tomcat v7. 0，单击下方的 Finish 按钮。

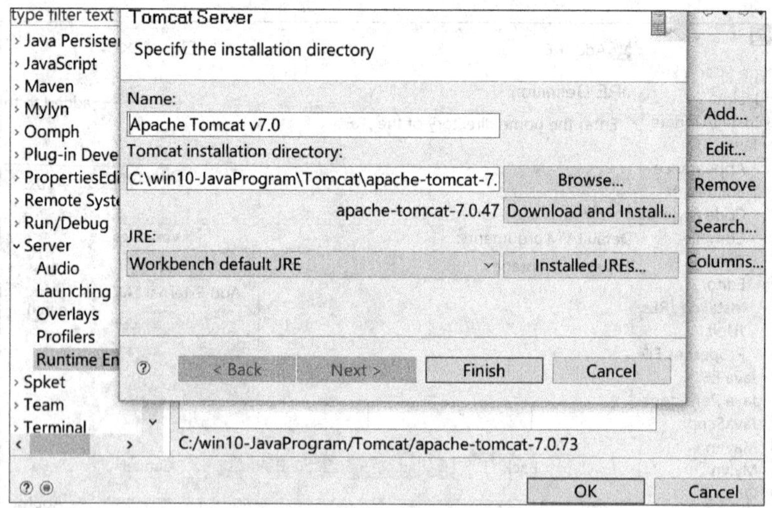

图 1-7　Tomcat Server 窗口

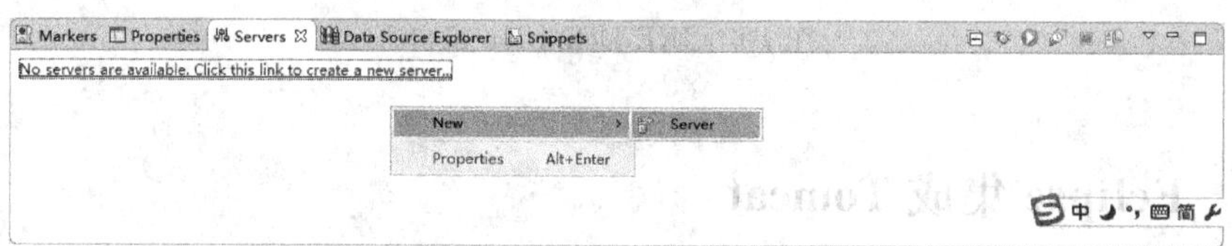

图 1-8　Eclipse 下方的 Servers 窗口

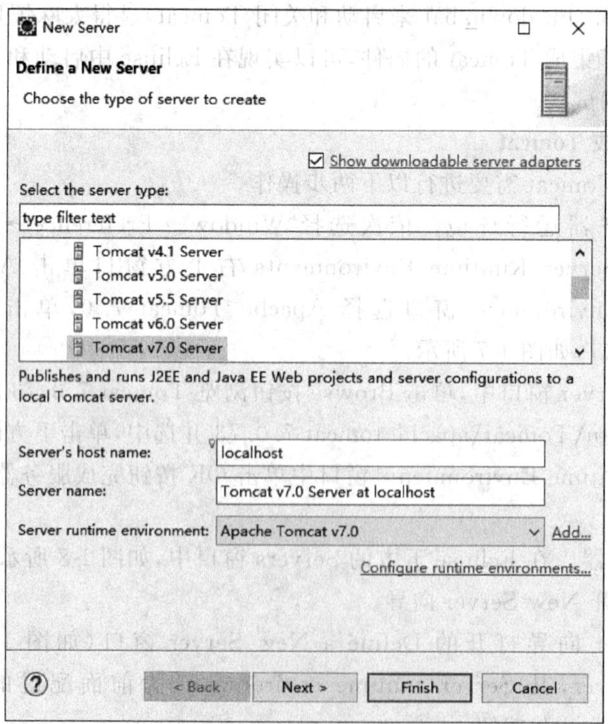

图 1-9　Define a New Server 窗口

完成上述工作后，在 Eclipse 左方的 Project Explorer 窗口中会显示 Servers 项目，如图 1-10 所示。单击 Servers，其下方出现 Tomcat 7 的配置情况，这样 Eclipse jee 集成 Tomcat 的工作就完成了。

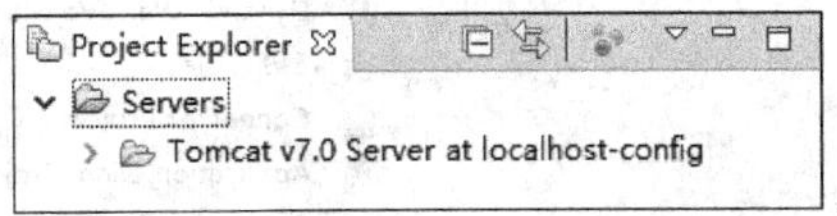

图 1-10　Project Explorer 窗口显示 Servers 项目

2. 测试 Eclipse 集成 Tomcat 是否成功

在 Eclipse 下方的 Servers 窗口中，单击工具栏中的 Start the server 按钮启动服务器 Tomcat，如图 1-11 所示。

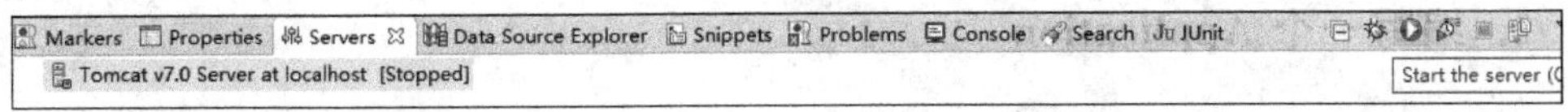

图 1-11　单击工具栏中的 Start the server 按钮启动服务器 Tomcat

这样就启动了前面安装和配置的服务器 Tomcat。在下方的 Console(控制台)窗口中显示 Tomcat 启动时的信息，如图 1-12 所示。

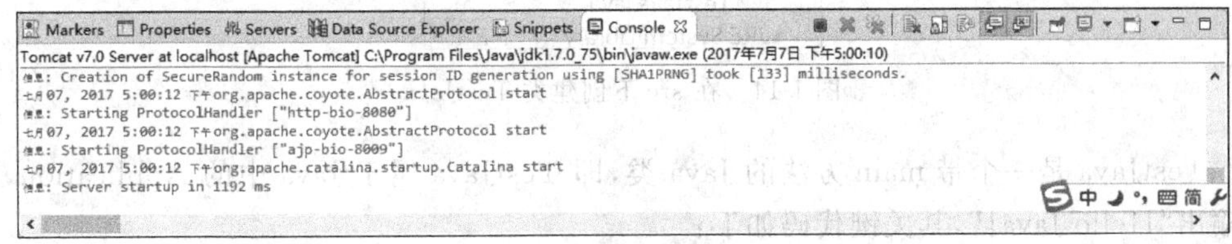

图 1-12　Console(控制台)窗口中显示 Tomcat 启动时的信息

1.2　Eclipse 基本使用

前面介绍了在 Eclipse 平台中编写 Java 程序必须首先创建项目。项目是 Eclipse 的最小工作单元。下面介绍在 Eclipse 中创建和运行 Java 项目、创建和运行 Java Web 项目的过程。

1.2.1　创建和运行 Java 项目

1. 创建 Java 项目

创建项目有多种方式，常见的是从 File 菜单创建项目。在 Eclipse 中，依次选择 File→New→Project 命令，如图 1-13 所示。

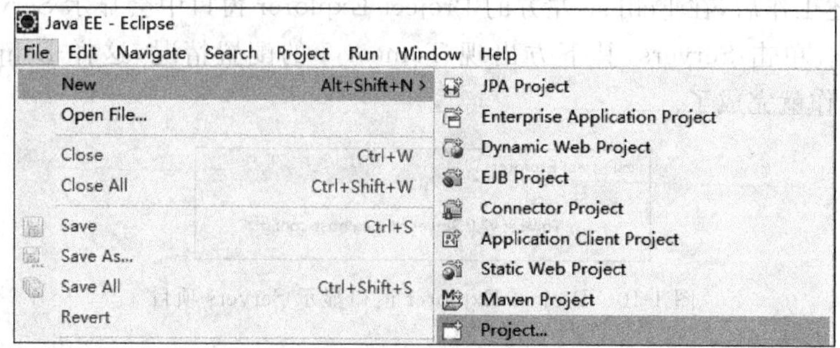

图 1-13　依次单击 File→New→Project 命令

在弹出的 New Project 窗口选择 Java Project，输入项目名称，如 ch01-01，单击下方的 Finish 按钮，完成 Java 项目 ch01-01 的创建。

2. 编写一个 Application

在项目 ch01-01 的 src 下创建一个类 TestJava，如图 1-14 所示。

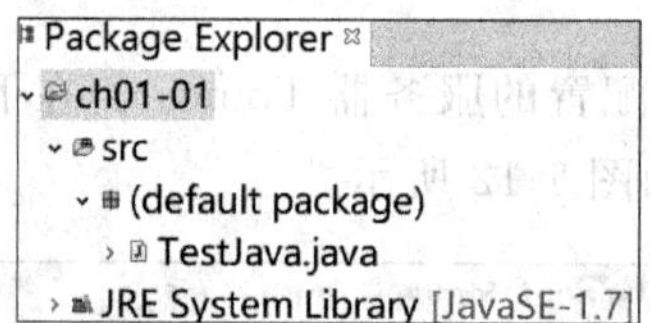

图 1-14　在 src 下创建类 TestJava

TestJava 是一个带 main 方法的 Java 类，即 TestJava 属于 Java 中的 Application，用于输出"Hello Java!"，其关键代码如下：

```
public class TestJava {
    public static void main(String[] args) {
        System.out.println("Hello Java!");
    }
}
```

3. 运行 Application

在 TestJava 类上右击，依次选择 Run As→Java Application 命令，如图 1-15 所示。

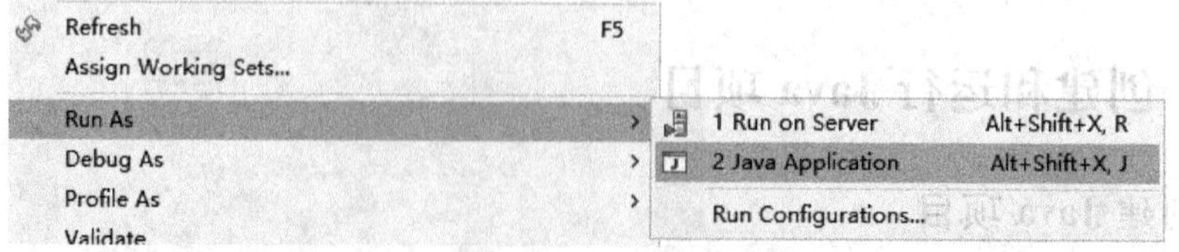

图 1-15　运行 Java Application

TestJava 运行后在控制台中的输出效果如图 1-16 所示。

```
Problems  Javadoc  Declaration  Console
<terminated> TestJava (1) [Java Application] C:\Program Files\Java\jdk1.7.0_75\bin\javaw.exe (2018年1月8日 上午9:48:16)
Hello Java!
```

图 1-16　TestJava 运行后在控制台输出效果

1.2.2　创建和运行 Java Web 项目

1. 创建 Java Web 项目

在 Eclipse 中，可以通过菜单 File→New→Dynamic Web Project 命令来创建一个 Java Web 项目，如图 1-17 所示。

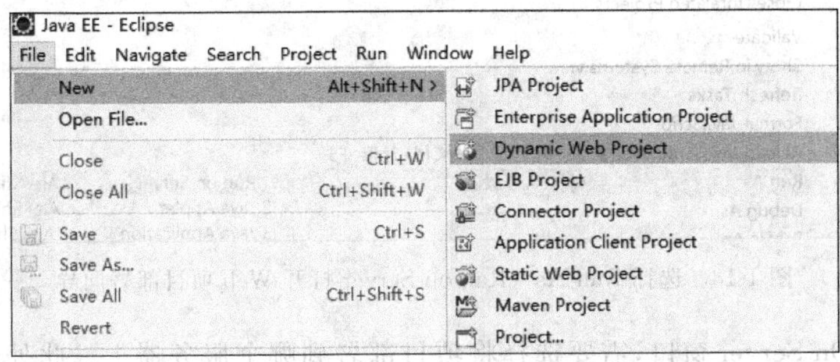

图 1-17　通过菜单 File→New→Dynamic Web Project 创建 Web 项目

在弹出的 New Web Project 窗口输入项目名称 ch01-02，单击下方的 Finish 按钮，完成创建 Java Web 项目的工作。

同样，Web 项目的 src 用于存放 Java 类，如 Servlet、JavaBean 等。但是 Web 资源文件如 JSP、HTML 页面、图片文件等需要在 Eclipse Web 项目的 WebContent 下创建，或者在 WebContent 下建立文件夹，分别存放各种资源文件。

例如，在 WebContent 下创建 index.jsp，其关键代码如下：

```
<%@page language="java" contentType="text/html; charset=UTF-8"
  pageEncoding="UTF-8"%>
<html><head><title>Insert title here</title></head>
<body>
  Hello Java Web!
</body>
</html>
```

2. 部署 Java Web 项目到 Tomcat 并运行

Java Web 项目需要部署到 Servlet 容器上，由 Servlet 容器执行。下面介绍如何将 Java Web 项目 ch01-02 部署到 Tomcat。

在 Projects Explorer 窗口，选中项目 ch01-02，右击选择 Run As→Run on Server 命

令，打开 Web 项目部署向导，如图 1-18 所示。

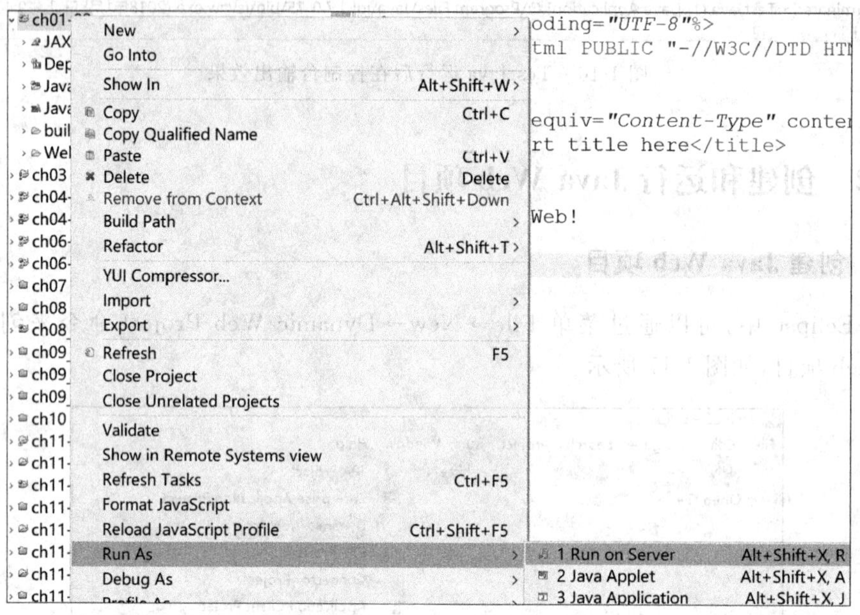

图 1-18　选择 Run As→Run on Server 打开 Web 项目部署向导

在 Run On Server 窗口，需要选择将项目部署到哪个服务器上。此处选择 Tomcat v7.0 Server at localhost，将项目 ch01-02 部署到 Tomcat 7 上，如图 1-19 所示。单击 Finish 按钮，Eclipse 会将 Web 项目部署到 Tomcat 7 上并启动 Tomcat 7 服务器。通常 Eclipse 还会打开内置浏览器显示 Web 项目首页 index.jsp 的效果，如图 1-20 所示。

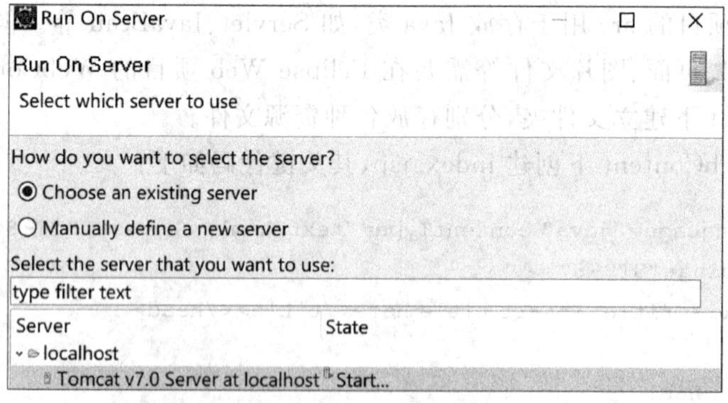

图 1-19　选择将项目部署到 Tomcat 7

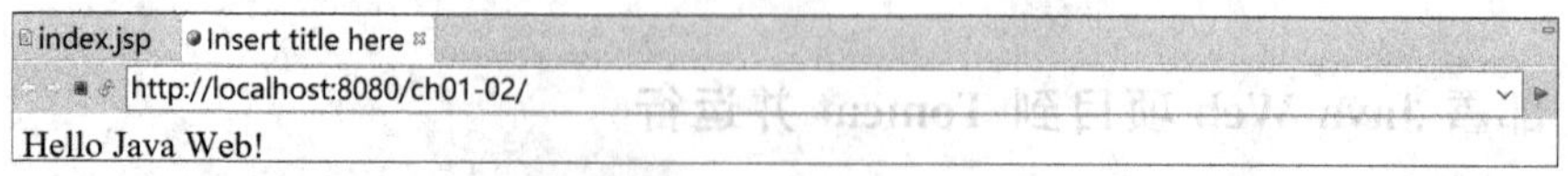

图 1-20　Eclipse 打开内置浏览器显示 Web 项目首页 index.jsp 的效果

小　　结

本章介绍了Java Web开发环境搭建和Eclipse的基本使用。Java Web开发环境搭建涉及JDK、Web服务器Tomcat、集成开发环境Eclipse、数据库服务器MySQL等相关软件的安装、配置、测试。Eclipse的基本使用包括创建和运行Java项目、创建和运行Java Web项目。

思考与习题

1. 请写出Java Web集成开发环境搭建需要的软件。
2. 请写出安装、配置、测试JDK的步骤。
3. Tomcat官方网站提供的两种类型的文件分别是什么？
4. 请写出安装、测试Tomcat的步骤。
5. 安装数据库服务器MySQL时的注意事项是什么？
6. Eclipse的工作空间的作用是什么？请写出配置Eclipse工作空间的步骤。
7. 请写出设置Eclipse的默认JRE步骤。
8. 请写出创建Eclipse Java项目的步骤。
9. 请写出运行一个Eclipse Java项目中的Application的步骤。
10. 请写出创建Eclipse Java Web项目的步骤。
11. 请写出运行一个Eclipse Java Web项目的步骤。

第2章

JDBC

应用程序通常需要与数据库交互,将数据保存到数据库,从数据库取出数据等。应用程序对数据库的操作主要有 4 种:插入记录、查询符合条件的记录、更新记录、删除记录,这 4 种操作常称为 CRUD(Create、Read、Update、Delete)。现在主流的数据库是关系数据库。常见的关系数据库有 Oracle、DB2、Microsoft SQLServer、MySQL 等。

在 Java 应用中访问数据库需要使用 JDBC。

2.1 JDBC 简介

2.1.1 JDBC 的概念

JDBC(Java Database Connectivity)是 Java 程序如何操作数据库的规范。在 Java 编写的程序中,程序员通过使用 JDBC API,可以用标准 SQL 语句访问几乎任何一种数据库;特定数据库厂商通过实现 JDBC API 生产的 JDBC 驱动程序完成某个具体数据库的实际操作。

JDBC 分为面向程序开发人员的 JDBC API 和面向底层 JDBC 驱动的 API。面向程序开发人员的 JDBC API 是学习的重点,面向底层 JDBC 驱动的 API 是数据库厂商实现驱动程序的规范,在此不予介绍。JDBC 体系结构如图 2-1 所示。

在 Java 应用程序中使用 JDBC API 来访问数据库时要在 classpath 中加载某个具体数据库的 JDBC 驱动,这样不管是访问什么数据库,只要有对应的数据库 JDBC 驱动,在 Java 程序中使用统一的类和接口就能完成对数据库的操作了。

2.1.2 JDBC 3.0 API 简介

JDBC 3.0 版本中,JDBC API 由 java.sql 和 javax.sql 包中的接口和类组成。

java.sql 包中的接口和类主要完成基本数据库操作,如创建数据库连接对象、创建执行 SQL 语句 Statement 对象或 PreparedStatement 对象、创建结果集对象等;同时也有一些高级数据库操作的处理,如批处理更新、事务隔离、可滚动结果集等。

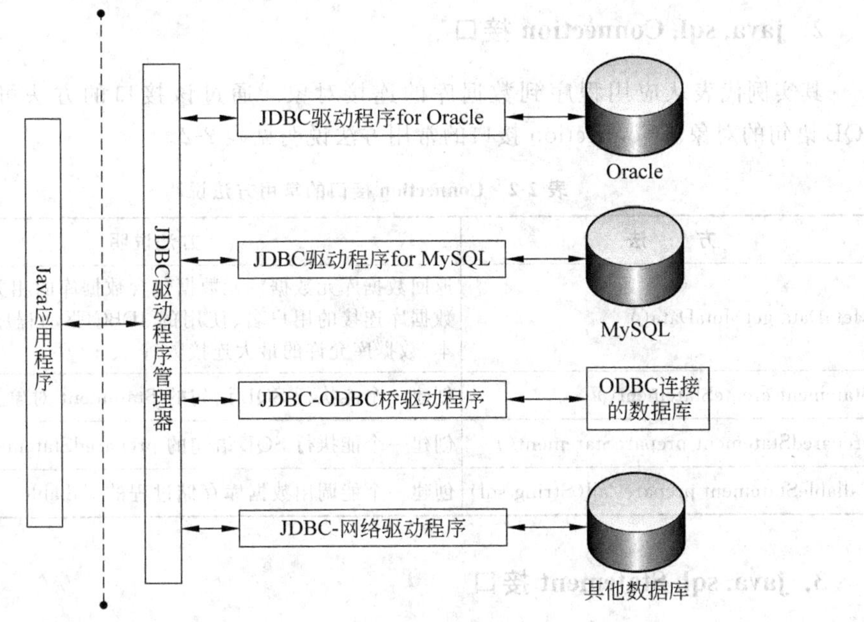

图 2-1　JDBC 体系结构

javax. sql 包中的类和接口主要完成高级数据库操作，如为连接管理、分布式事务和旧有的连接提供更好的抽象，它引入了容器管理的连接池、分布式事务和行集等。

JDBC API 常用的类和接口如图 2-2 所示。

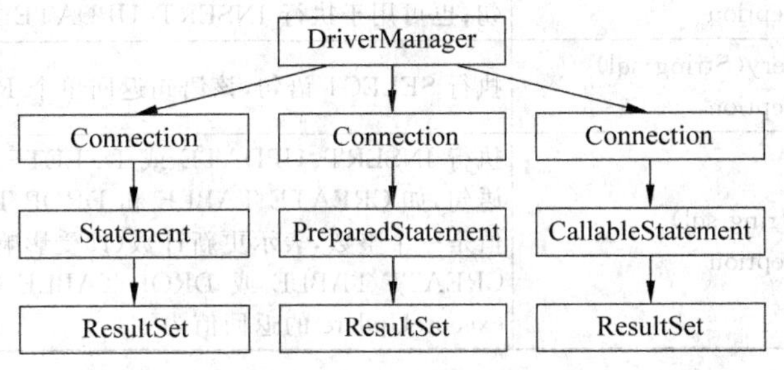

图 2-2　JDBC 常用的类和接口

1. java. sql. DriverManager 类

该类管理注册的驱动程序，获得从应用程序到数据库的连接对象。DriverManager 类的常用方法说明见表 2-1。

表 2-1　DriverManager 类的常用方法说明

方　　法	方法说明
getConnection(String url, Sting user, String pwd)	返回 Connection 对象

2. java.sql.Connection 接口

其实例代表从应用程序到数据库的连接对象。通过该接口的方法可以获得执行 SQL 语句的对象。Connection 接口的常用方法说明见表 2-2。

表 2-2 Connection 接口的常用方法说明

方　法	方法说明
MetaData getMetaData()	返回数据库元数据。元数据包含数据库的相关信息，如当前数据库连接的用户名、使用的 JDBC 驱动程序、数据库的版本、数据库允许的最大连接数等
Statement createStatement()	创建一个能执行 SQL 语句的 Statement 对象
preparedStatement prepareStatement()	创建一个能执行 SQL 语句的 preparedStatement 对象
CallableStatement prepareCall(String sql)	创建一个能调用数据库存储过程的 CallableStatement 对象

3. java.sql.Statement 接口

其实例用于执行静态 SQL 语句。Statement 接口的常用方法说明见表 2-3。

表 2-3 Statement 接口的常用方法说明

方　法	方法说明
boolean execute(String sql) throws SQLException	用于执行返回多个结果集、多个更新计数或二者组合的语句；也可用于执行 INSERT、UPDATE 或 DELETE 语句
ResultSet executeQuery(String sql) throws SQLException	执行 SELECT 语句，该语句返回单个 ResultSet 对象
int executeUpdate(String sql) throws SQLException	执行 INSERT、UPDATE 或 DELETE 语句或者 SQL DDL 语句，如 CREATE TABLE 和 DROP TABLE。该方法返回值是一个整数，表示更新计数(即受影响的记录行数)。对于 CREATE TABLE 或 DROP TABLE 等不操作行的语句，executeUpdate 的返回值为零

4. java.sql.PreparedStatement 接口

PreparedStatement 是 Statement 的子接口，用于执行动态的 SQL 语句。PreparedStatement 接口的使用见 2.4.2 节。

5. java.sql.CallableStatement 接口

CallableStatement 是 PreparedStatement 的子接口，用于调用数据库存储过程。CallableStatement 的使用见 2.4.3 节。

6. java.sql.ResultSet 接口

其实例表示执行 SQL 查询后的结果集对象。ResultSet 接口的常用方法说明见

表 2-4。

表 2-4　ResultSet 接口的常用方法说明

方　法	方法描述
boolean next() throws SQLException	将光标从结果集的当前位置下移一行。ResultSet 光标最初位于第一行之前。第一次调用 next() 方法光标指向第一行；第二次调用 next() 方法光标指向第二行，以此类推。若光标指向的行有记录存在，则该方法返回 true，否则返回 false
int getInt(int columnIndex) throws SQLException	返回结果集中 columnIndex 位置的字段的值，该字段对应的 Java 类型是 int
int getInt(String columnName) throws SQLException	返回结果集中字段 columnName 的值，该字段对应的 Java 类型是 int
StringgetString(int columnIndex) throws SQLException	返回结果集中 columnIndex 位置的字段的值，该字段对应的 Java 类型是 String
StringgetString(String columnName) throws SQLException	返回结果集中字段 columnName 的值，该字段对应的 Java 类型是 String

ResultSet 的其他 getter 方法还有 getFloat()、getDouble()、getBytes()、getLong()、getBoolean()、getClob()、getBlob()等。

2.1.3　JDBC 驱动程序类型

JDBC 驱动程序有下列 4 种类型。

1. JDBC-ODBC 桥驱动程序

早期 Java 刚出现的时候，许多数据库没有 JDBC 驱动但是有 ODBC 驱动。为了实现 Java 程序使用标准 SQL 语句访问数据库，必须借助于 JDBC-ODBC 桥这种驱动程序，将 JDBC 调用转换为 ODBC 调用，再通过 ODBC 访问数据库。因此在访问数据库的每个客户端都必须安装 ODBC 驱动程序。JDBC-ODBC 桥驱动程序的缺点是增加了 ODBC 层后导致效率低。现在主流数据库都有 JDBC 驱动，因此该方式很少使用，除非是访问没有 JDBC 驱动的数据库(如 Microsoft 的 Access)。使用该驱动程序访问数据库如图 2-3 所示。

2. 部分 Java 驱动程序

大部分数据库厂商提供访问数据库的 API 由 C 或其他语言编写，依赖具体的平台。部分 Java 驱动程序由 Java 编写，它调用数据库厂商提供的本地 API。在应用程序中使用 JDBC API 访问数据库时，JDBC 驱动程序将 JDBC 调用转换成本地 API 调用；数据库处理完请求，将结果通过本地 API 返回，进而返回给 JDBC 驱动程序，JDBC 驱动程序将结果转换成 JDBC 标准形式，再返回给应用程序。部分 Java 驱动程序直接将 JDBC API 翻

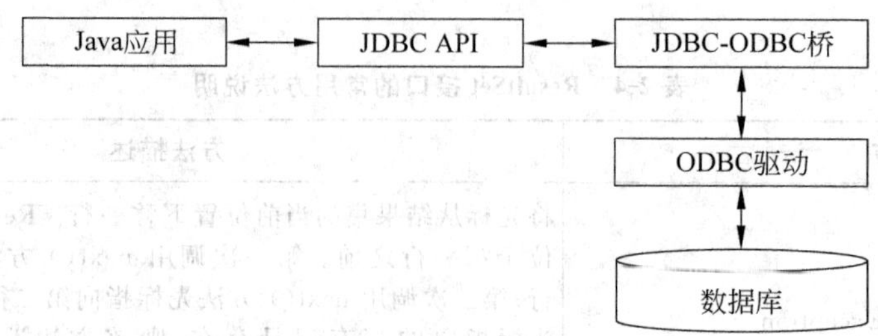

图 2-3　使用 JDBC-ODBC 桥访问数据库

译成具体数据库的 API,因此效率比第一种驱动高。它的缺点是客户端需要安装特定数据库的驱动。使用部分 Java 驱动程序访问数据库如图 2-4 所示。

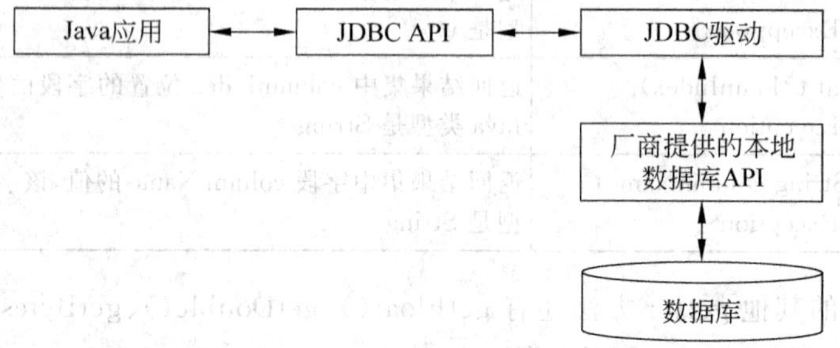

图 2-4　使用部分 Java 驱动程序访问数据库

3. 中间件驱动程序

这种中间件驱动程序属于纯 Java 驱动程序,它将 JDBC API 转换成独立于数据库的协议。中间件驱动程序并不是直接与数据库进行通信而是与一个中间件服务器通信,然后这个中间件服务器和数据库进行通信。这种额外的中间层次提供了灵活性:可以用相同的代码访问不同的数据库。另外中间件服务器可以安装在专门的硬件平台上进行优化。BEA Weblogic Server 使用该方式进行数据库访问。这种驱动程序的缺点是额外的中间件层次可能降低整体系统性能。使用中间件驱动程序访问数据库如图 2-5 所示。

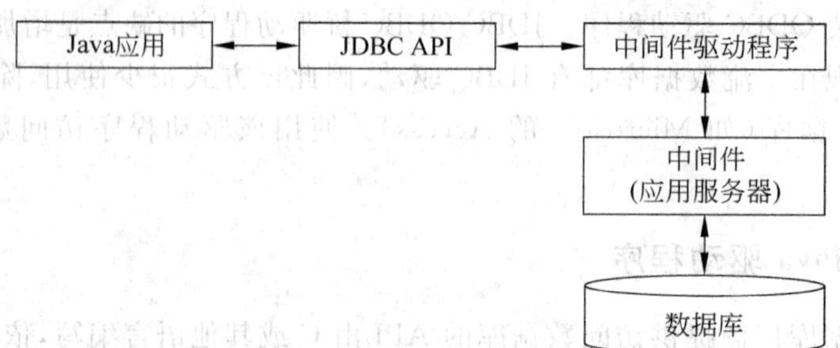

图 2-5　使用数据库中间件驱动程序访问数据库

4. 纯 Java 驱动程序

这种驱动程序直接与数据库进行通信,因此性能最好;另外可以利用数据库提供的特殊功能。这种驱动程序本质是使用 Socket 编程。目前几乎所有的数据库都提供纯 Java 驱动程序。建议使用该类驱动程序。本书使用 JDBC 访问数据库 MySQL 的案例都是使用该类型的驱动程序。使用纯 Java 驱动程序访问数据库如图 2-6 所示。

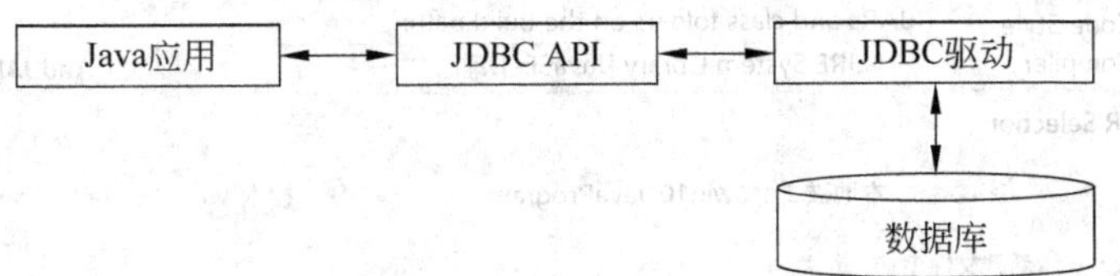

图 2-6　使用纯 Java 驱动程序访问数据库

2.2　使用 JDBC

2.2.1　准备工作

在 Java 程序中使用 JDBC 访问数据库之前,需要进行准备工作,包括创建数据库和表、创建 Eclipse Java 项目、配置项目构建路径添加 JDBC 驱动。

1. 创建数据库和表

安装 MySQL 数据库服务器后,新建一个数据库 testDB 和表 user。表 user 对应的 DDL 如下:

```
CREATE TABLE `user` (
  `id` int(11) NOT NULL AUTO_INCREMENT,
  `name` varchar(50) DEFAULT NULL,
  `password` varchar(50) DEFAULT NULL,
  `sex` varchar(10) DEFAULT NULL,
  `birthday` datetime DEFAULT NULL,
  PRIMARY KEY (`id`)
) ENGINE=InnoDB AUTO_INCREMENT=1 DEFAULT CHARSET=utf8;
```

2. 创建 Eclipse Java 项目

在 Eclipse 中创建 Java 项目 ch02。

3. 配置项目构建路径添加 JDBC 驱动程序

在 Eclipse 中,右击项目 ch02,依次选择 Build Path→Configure Build Path 命令,打

开构建路径配置对话框，选择右方窗口 Java Build Path 下的 Libraries 标签，单击 Add JARs 按钮，浏览 MySQL 的 JDBC 驱动程序 mysql-connector-java-5.0.6-bin.jar 所在目录，并选择 mysql-connector-java-5.0.6-bin.jar，如图 2-7 所示。

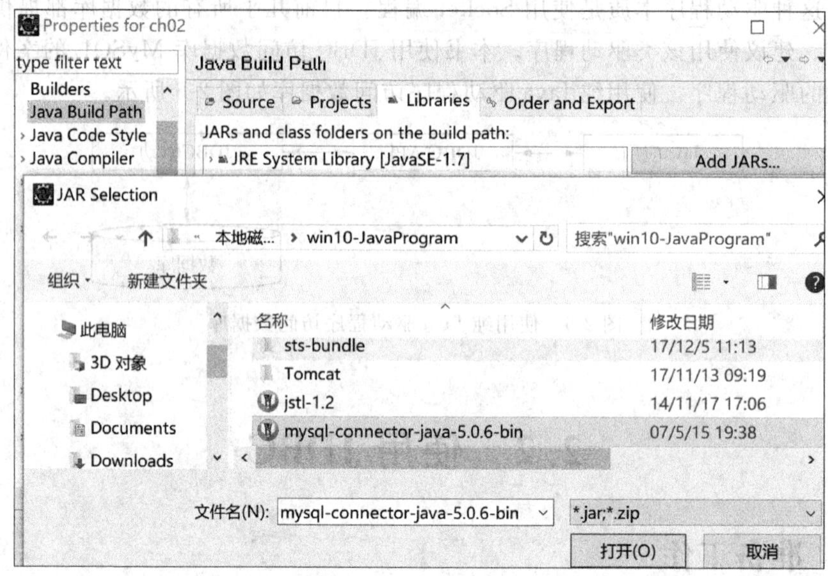

图 2-7 定位和选择 mysql-connector-java-5.0.6-bin.jar

单击图 2-7 中的"打开"按钮。最后单击 Java Build Path 窗口下方的 OK 按钮结束配置项目构建路径添加 JDBC 驱动的工作。

添加 mysql-connector-java-5.0.6-bin.jar 后项目 ch02 的结构如图 2-8 所示。

ch02
src
JRE System Library [JavaSE-1.7]
mysql-connector-java-5.0.6-bin.jar

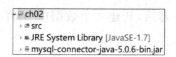

图 2-8 添加 mysql-connector-java-5.0.6-bin.jar 后项目 ch02 的结构

2.2.2 使用 JDBC 一般步骤

1. 加载驱动程序

使用 Class.forName()方法显式地加载数据库 MySQL 5 的 JDBC 驱动程序：

```
Class.forName("com.mysql.jdbc.Driver");
```

该方法的定义如下：

```
public static Class forName(String className) throws ClassNotFoundException
```

该方法加载参数 className 指定的类文件；如果找不到 className 指定的类文件，该方法会抛出 ClassNotFoundException。

2. 创建连接对象

加载后的驱动程序由 java. sql. DriverManager 类的实例管理。该类有重载的 getConnection()方法能得到从应用程序到数据库的连接对象，即 Connection 类型的对象。

创建 MySQL 5 数据库连接的关键代码如下：

```
Connection conn =
    DriverManager.getConnection("jdbc:mysql://localhost:3306/testDB ","root",
"root");
```

DriverManager 的 getConnection 方法定义如下：

```
ConnectiongetConnection(url,user,password)
```

其中 url 是数据库的网络位置，user 和 password 是访问数据库的用户名和密码。

URL 的一般形式如下：

```
jdbc:<subprotocol>://<datasourcename>
```

- jdbc 表示 Java 程序连接数据库的协议是 jdbc；目前只能是 jdbc 协议。
- subprotocol(子协议)主要用于识别数据库驱动程序，即不同数据库驱动程序的子协议不同。
- datasourcename(数据源名)包括数据库的 IP 地址、端口号、数据库名称。

访问本机上的 MySQL 5 数据库 testDB 的 URL 为：

```
jdbc:mysql://localhost:3306/testDB
```

user 和 password 分别是登录 MySQL 5 的用户名及其密码。由于在安装 MySQL 5 的时候设置了 root 用户的密码为 root。因此，此处的 user 和 password 都是 root。

3. 创建 Statement 对象执行 SQL 语句

Connection 定义了创建 Statement 对象或 PreparedStatement 对象的方法。下面以创建 Statement 对象为例说明。

首先定义 String 类型的 SQL 语句：

```
String sql ="select id,name,password,sex,birthday from user";
```

接着调用 Connection 对象 conn 的 createStatement()方法获得 Statement 类型的对象 stmt：

```
Statement stmt =conn.createStatement();
```

使用 Statement 的 executeQuery(String sql)方法进行查询，得到查询数据库返回的结果集 ResultSet 对象。

```
ResultSet rs =stmt.executeQuery(sql);
```

如果 SQL 语句是 insert、delete、update 语句，那么应该调用 executeUpdate(sql)。例如如下代码会在 user 表中插入一条记录。

```
String sql="insert into user (name,password,sex) values('zhangsan', '123', 'man')";
Statement stmt=conn.createStatement();
stmt.executeUpdate(sql);
```

4. 遍历结果集对象(可选)

在使用 JDBC 访问数据库的第 3 步如果是执行 SQL 查询，该步骤对查询后的结果集对象进行遍历，通常可以用如下 while 循环实现。

```
String sql ="select * from user";
Statement stmt =conn.createStatement();
ResultSet rs =stmt.executeQuery(sql);
while (rs.next()) {
    int id =rs.getInt("id");
    String name =rs.getString("name");
    String password =rs.getString("password");
    String sex =rs.getString("sex");
    int age =rs.getInt("age");
    Date birthday =rs.getDate("birthday");
    System.out.println("id=" +id +";name=" +name +";password="
        +password +";sex=" +sex +";age=" +age +";birthday="
        +birthday);
}
```

ResultSet 表示从数据库进行 SQL 查询后得到的结果集。例如，数据库 testDB 中表 user 的记录如表 2-5 所示。

表 2-5　数据库 testDB 中表 user 的记录

id	name	password	sex	birthday
1	zhangsan	123	man	1990-12-31
2	lisi	456	man	1989-12-1

执行如下 SQL 查询：

```
String sql="select id,name,password,sex,birthday from user";
Statement stmt=conn.createStatement();
ResultSet rs =stmt.executeQuery(sql);
```

结果集对象 rs 如表 2-6 所示。

表 2-6 中最左侧的→表示在 rs 中有个光标，每次调用 next()方法，光标会下移一行。

ResultSet.next()方法的返回值是 boolean 型，表示调用 next()方法后，如果光标指向的当前行记录存在，则 next()方法返回 true；否则返回 false。

表 2-6 执行查询后的结果集对象 rs

	id	name	password	sex	birthday
→					
	1	zhangsan	123	man	1990-12-31
	2	lisi	456	man	1989-12-1

注意：在执行了

```
ResultSet rs =stmt.executeQuery(sql);
```

之后，这时候，指针→指向查询结果集中第一条记录的前一行。此时第一次调用

```
rs.next();
```

那么指针→指向第一条记录所在的行，如表 2-7 所示。

表 2-7 第一次调用 rs.next()后指光标向第一行

	id	name	password	sex	birthday
→	1	zhangsan	123	man	1990-12-31
	2	lisi	456	man	1989-12-1

第二次调用 rs.next()方法，光标指向第二行，如表 2-8 所示。

表 2-8 第二次调用 rs.next()后光标指向第二行

	id	name	password	sex	birthday
	1	zhangsan	123	man	1990-12-31
→	2	lisi	456	man	1989-12-1

第三次调用 rs.next()方法，此时光标指向 rs 的最后，此行没有记录，因此 rs.next()方法返回 false，如表 2-9 所示。

表 2-9 第三次调用 rs.next()后光标指向 rs 的最后

	id	name	password	sex	birthday
	1	zhangsan	123	man	1990-12-31
	2	lisi	456	man	1989-12-1
→					

当调用 rs.next()方法后，如果光标指向的行有记录存在，那么可以通过 rs.getter()方法得到记录的某个字段的值。getter ()方法可以是 getString()、getInt()、getFloat()、getDate()等。

使用 getter ()方法获取记录的字段值有两种用法：根据查询 SQL 语句中字段名或查询 SQL 语句中字段的索引。

对于如下 SQL 语句：

```
String sql ="select id,name,password,sex,birthday from user";
```

可以使用

```
int id =rs.getInt("id");    //通过 SQL 语句中字段名来获取 id 字段的值
```

或者使用

```
int id =rs.getInt(1);       //通过 SQL 语句中字段索引来获取 id 字段的值
```

第一种用法较为常见，可以避免查询字段数目较多时根据索引获取字段值容易出错的问题。

使用 rs.getInt("id")是因为 id 字段在数据库中是 int(整型)类型，对应的 Java 类型是 int 或 Integer。

又如：

```
String name =rs.getString("name");
```

使用 rs.getString("name")是因为 name 字段在数据库中是 varchar(可变字符串)类型，对应的 Java 类型是 String。

MySQL 常用数据类型与 Java 类型的对应情况见表 2-10。

表 2-10 MySQL 常用数据类型与 Java 类型的对应情况

MySQL 数据类型	Java 类型
int	java.lang.Integer 或 int
float	java.lang.Float 或 float
double	java.lang.Double 或 double
char	java.lang.String
varchar	java.lang.String
date	java.sql.Date
time	java.sql.Time
datetime	java.sql.TimeStamp
timestamp	java.sql.Timestamp

5. 依次关闭资源对象

Connection 对象、Statement(或 PreparedStatement)对象、ResultSet 对象均是占用资源的对象，使用完毕后需要关闭这些资源对象。这些对象都有 close()方法表示关闭资源对象本身。资源对象的关闭顺序与创建顺序相反。例如如下代码表示依次关闭 rs、stmt、conn 对象。

```
rs.close();
stmt.close();
conn.close();
```

2.3 JDBC 实现增、删、改、查

本节介绍实现数据库常见的 4 种操作，即实现插入记录、查询记录、更新记录和删除记录。创建 CRUDTest 类，该类有 add()、listAll ()、update()、delete()方法，分别实现数据库 testDB 中表 user 插入记录、查询、更新和删除记录的操作。在该类的 main()方法中分别调用 add()、listAll()、update()、delete()方法。

2.3.1 插入记录

首先，在 Eclipse 的 Java 项目 ch02 下创建 CRUDTest 类。该类带有 main()方法，在 Java 中属于 Application。

接着，编写实现数据库插入记录的 add()方法的关键代码如下：

```
...
public class CRUDTest {
    public static void add() throws ClassNotFoundException, SQLException {
        //定义创建数据库连接所需的参数
        String url ="jdbc:mysql://localhost:3306/testDB";
        String dbUser ="root";
        String dbPassword ="root";
        String driverName ="com.mysql.jdbc.Driver";
        //1. 加载驱动程序
        Class.forName(driverName);
        //2. 创建连接对象
        Connection conn =DriverManager.getConnection(url, dbUser,
            dbPassword);
        //3. 创建 Statement 对象执行 SQL 语句
        //id 为数据库自增型，因此不需要显式插入 id
        String sql ="insert into user(name,password,sex,,birthday) "
            +"values('zhangsan','123','man','1990-12-31')";
        Statement stmt =conn.createStatement();
        int lines =stmt.executeUpdate(sql);
        //4. 遍历结果集(此处不需要)

        System.out.println("lines=" +lines); //输出受影响的记录行数
        //5. 关闭资源对象
        stmt.close();
        conn.close();
    }
}
```

在以上代码中，加载驱动程序 Class. forName（driverName）可能会抛出 ClassNotFoundException；创建连接对象、创建语句对象、执行 SQL 语句等方法可能会抛出 SQLException。此处采用了 Java 异常处理机制中向上抛出异常的方法，即在 add()方法内部不捕获和处理异常，而是谁调用 add()方法，谁负责捕获和处理异常。

```
int lines =stmt.executeUpdate(sql);
System.out.println("lines=" +lines);
```

以上两行代码将执行 SQL 后的数据库受影响的记录行数赋值给 lines，并打印输出 lines。

最后，在 CRUDTest 类中增加 main()方法，在 main()方法中调用 add()方法。CRUDTest 类关键代码如下：

```
…
public class CRUDTest {
    public static void main(String[] args) throws ClassNotFoundException,
        SQLException {
        add();
    }
    public static void add() throws ClassNotFoundException, SQLException {
            //… add()方法代码同上,此处略
    }
}
```

在 main()方法中也没有捕获和处理异常，仍旧向上抛出异常；异常会由调用 main()方法的 Java 虚拟机捕获。

调用 add()方法后 Eclipse 控制台输出效果如图 2-9 所示。

```
Markers  Properties  Servers  Data Source Explorer  Snippets  Problems  Console ☒  Search
<terminated> CRUDTest [Java Application] C:\Program Files\Java\jdk1.7.0_75\bin\javaw.exe (2017年7月9日 上午11:22:07)
lines=1
```

图 2-9　调用 add()方法后 Eclipse 控制台输出效果

可以使用 MySQL 客户端工具查看 testDB 数据库的 user 表，确保记录插入成功。

2.3.2　查询记录

listAll()方法实现查询 user 表的全部记录，其关键代码如下：

```
public static void listAll() throws ClassNotFoundException, SQLException {
…
    Class.forName(driverName);
    Connection conn =DriverManager.getConnection(url, dbUser,
        dbPassword);
```

```
        String sql ="select * from user";
        Statement stmt =conn.createStatement();
        ResultSet rs =stmt.executeQuery(sql);
        while (rs.next()) {
            int id =rs.getInt("id");
            String name =rs.getString("name");
            String password =rs.getString("password");
            String sex =rs.getString("sex");
            Date birthday =rs.getDate("birthday");
            System.out.println("id=" +id +";name=" +name +";password="
                  +password +";sex=" +sex +";birthday="+birthday);
        }
        rs.close();
        stmt.close();
        conn.close();
    }
```

遍历结果集时使用输出语句 System. out. println()将每条记录输出到控制台。

在 CRUDTest 类的 main()方法中,将 add()方法注释掉,并增加对 listAll()方法的调用,关键代码如下:

```
...
public class CRUDTest {
    public static void main(String[] args) throws ClassNotFoundException,
    SQLException {
        //add();
        listAll();
    }
    public static void add() throws ClassNotFoundException,
                SQLException {
        ...
    }
    public static void listAll() throws ClassNotFoundException,
                SQLException {
        ...
    }
}
```

2.3.3 更新记录

update()方法实现将 user 表中 id=1 的记录的 password 字段值更新为 666,其关键代码如下:

```
public static void update() throws ClassNotFoundException, SQLException {
```

```
    …
    Class.forName(driverName);
    Connection conn =DriverManager.getConnection(url, dbUser, dbPassword);
    String sql ="update user set password=`666` where id=1";
    Statement stmt =conn.createStatement();
    int lines =stmt.executeUpdate(sql);
    System.out.println("lines=" +lines);
    stmt.close();
    conn.close();
}
```

在 CRUDTest 类的 main()方法中，将 listAll()等方法注释掉，并增加对 update()方法的调用，关键代码如下：

```
…
public class CRUDTest {
    public static void main(String[] args) throws ClassNotFoundException,
        SQLException {
        //add();
        //listAll();
        update();
    }
    …
    public static void update() throws ClassNotFoundException,
        SQLException {
        //update()方法代码略
    }
}
```

2.3.4 删除记录

delete()方法实现删除 id=1 的记录，其关键代码如下：

```
public static void delete() throws ClassNotFoundException, SQLException {
    …
    Class.forName(driverName);
    Connection conn =DriverManager.getConnection(url, dbUser, dbPassword);
    String sql ="delete from user where id=1";
    Statement stmt =conn.createStatement();
    int lines =stmt.executeUpdate(sql);
    System.out.println("lines=" +lines);
    stmt.close();
    conn.close();
}
```

在 CRUDTest 类的 main()方法中，将 update()等方法注释掉，并增加对 delete()方法的调用，代码如下所示：

```
public class CRUDTest {
    public static void main(String[] args) throws ClassNotFoundException,
        SQLException {
        //add();
        //listAll();
        //update();
        delete();
    }
    …
    public static void delete() throws ClassNotFoundException, SQLException {
        //delete()方法代码略
    }
}
```

2.4 JDBC 进阶

2.4.1 数据库帮助类 DBConnection

在前面介绍的实现数据库插入、查询、更新、删除记录的 CRUDTest 类程序代码中，有 4 个方法来实现 CRUD 操作；但是发现在每个方法中都存在加载驱动程序、获得连接对象、关闭资源对象等重复代码。

在实际程序开发中，常常创建一个用于获得数据库连接的帮助类 DBConnection，这样就避免了代码重复，实现了代码重用和面向对象中的类设计的高内聚。DBConnection 的关键代码如下：

```
package util;
…
public class DBConnection {
    private static final String driverName ="com.mysql.jdbc.Driver";
    private static final String url ="jdbc:mysql://localhost:3306/testDB";
    private static final String user ="root";
    private static final String password ="root";
    private DBConnection() {}
    static {
        try {
            Class.forName(driverName);
        } catch (ClassNotFoundException e) {
            e.printStackTrace();
```

```
    }
  }

  public static Connection getConnection() throws SQLException {
      return DriverManager.getConnection(url, user, password);
  }

  public static void close(ResultSet rs, Statement st, Connection conn) {
      try {
          if (rs !=null) {
              rs.close();
          }
      } catch (SQLException e) {
          e.printStackTrace();
      } finally {
          try {
              if (st !=null) {
                  st.close();
              }
          } catch (SQLException e) {
              e.printStackTrace();
          } finally {
              if (conn !=null) {
                  try {
                      conn.close();
                  } catch (SQLException e) {
                      e.printStackTrace();
                  }
              }
          }
      }
  }
}
```

使用 DBConnection 类后，插入记录的 add()方法的关键代码如下：

```
...
public class CRUDTestByDBConnection {
    public static void main(String[] args)
        throws ClassNotFoundException,SQLException {
        add();
        // get();
        // update();
        // delete();
    }
```

```
    public static void add() throws SQLException {
        Connection conn =null;
        Statement st =null;
        ResultSet rs =null;
        try {
            // 1. 获得连接对象
            conn =DBConnection.getConnection();
            // 定义 SQL 语句
            String sql ="insert into user(id,name,password,sex,age,birthday)"
                    +"values(1,'zhangsan','123','man',22,'1990-12-31')";
            // 2. 创建语句对象
            st =conn.createStatement();
            st.executeUpdate(sql);
            // 3. 遍历结果集(此处不需要)
        } finally {
            // 4. 关闭资源对象
            DBConnection.close(rs, st, conn);
        }
    }
}
```

2.4.2 使用 PreparedStatement

java.sql.PreparedStatement 支持预编译的 SQL 语句。如果多次访问数据库的 SQL 语句只是参数不同时，那么该对象比 Statement 对象的效率高。另外，使用 PreparedStatement 对象可以避免 SQL 注入问题。

PreparedStatement 接口是 Statement 接口的子接口，它的 execute()方法、executeQuery()方法、executeUpdate()方法的含义与 Statement 接口的这几个同名方法一样。

使用 PreparedStatement 时与 Statement 的不同之处有 3 点：

(1) 创建 PreparedStatement 对象时需要 SQL 语句；

(2) 由于 SQL 语句中含有参数，因此需要使用 PreparedStatement 的 setXxx 方法为参数赋值；

(3) 执行 SQL 语句的 execute()方法、executeQuery()方法、executeUpdate()方法不需要 SQL 语句。

下面以第 1 章的插入记录为例，介绍 PreparedStatement 的使用。

```
//1. 加载驱动(代码略)；
//2. 创建连接对象(代码略)；
//定义 SQL 语句；
```

```
String sql="insert into user(name,password,sex,birthday) values(?,?,?,?)";
//3.创建 PreparedStatement 的对象 ps(需要 SQL 语句)
PreparedStatement ps=conn.PreparedStatement(sql);
//需要调用 setXxx()方法设置参数
ps.setString(1,"zhangsan");
ps.setString(2,"123");
ps.setString(3,"man");
ps.setDate(4,"1990-12-31");
//执行 SQL 语句时的方法不带参数
ps.executeUpdate();
//4.遍历结果集(此处不需要)
//5.关闭资源对象
ps.close();
conn.close();
```

在向 user 表插入一条记录的 SQL 语句中的字段的值是动态的，用参数？表示。setXxx()方法为参数？赋值。在为？赋值时，如果？对应的字段在数据库是 varchar 或 char 等字符类型就使用 ps.setString()；该方法的第一个参数表示在当前的 SQL 语句中参数？的索引(从 1 开始)，第二个参数表示替代参数？的实际值。

例如，"ps.setString(2, "123");"表示将用实际值 123 作为该用户的密码替代 SQL 语句中的第二个?，因为在 SQL 语句中第二个？对应 password 字段。其他的类推即可。

"ps.executeUpdate();"表示执行 SQL 语句。PreparedStatement.executeUpdate()与 Statement.executeUpdate(sql)不同，不能带参数 sql 语句，否则程序会出错。

PreparedStatement 接口的其他常用方法说明见表 2-11。

表 2-11　PreparedStatement 接口的其他常用方法说明

方　法	方法说明
void setInt(int parameterIndex,int x) throws SQLException	这些方法称为 setter 方法；给 parameterIndex 位置的参数赋值。 参数类型可以是 int、float、double、long、boolean、string
void setString(int parameterIndex,string x) throws SQLException	
void setFloat(int parameterIndex, float x) throws SQLException	

2.4.3　调用存储过程

1. 存储过程的概念

存储过程是一个 SQL 语句和可选控制流语句的预编译集合。存储过程编译完成后存放在数据库中，这样就省去了执行 SQL 语句时对 SQL 语句进行编译所花费的时间。在执行存储过程时只需要将参数传递到数据库中，而不需要将整条 SQL 语句都提交给数据库,从而减少了网络传输的流量,从另一方面提高了程序的运行速度。

2. 创建存储过程的语法格式

创建存储过程的语法格式如下:

```
create procedure 存储过程名([[IN |OUT |INOUT ] 参数名   数据类型...])
begin
    存储过程体
end
```

3. JDBC 调用存储过程的一般步骤

JDBC API 通过 java. sql. CallableStatement 接口的对象调用存储过程。CallableStatement 继承自 PreparedStatement 接口。

JDBC 调用存储过程的一般步骤如下:

(1) 定义调用存储过程的 SQL 语句。

```
String sql ="{call <procedure-name>(?,?,...)}";
```

其中,?,?,... 表示存储过程的参数列表。

(2) 获取 CallableStatement 对象。

```
CallableStatement cs =conn.prepareCall(sql); //conn 是 Connection 类型的对象
```

(3) 设置参数(可选)。

如果存储过程带有输入参数,那么需要使用 CallableStatement 对象的 setter()方法对参数赋值。例如,“cs. setString(1, "callableStatement");”表示为存储过程语句中的第一个? 赋值为 String 类型的"callableStatement"。

如果存储过程带有输出参数,那么需要使用 CallableStatement 对象的 registerOutParameter()方法将输出参数注册为 JDBC 类型。例如,“cs. registerOutParameter(1, Types. VARCHAR);”表示将存储过程输出参数注册为 VARCHAR 类型。java. sql. Types 定义了 JDBC 类型。

如果存储过程带有输入输出参数,那么需要使用 CallableStatement 对象的 setter()方法对参数赋值,再使用 CallableStatement 对象的 registerOutParameter()方法将输出参数注册为 JDBC 类型。例如,以下语句首先使用 setInt()方法为存储过程输入输出参数赋值为 1,再使用 registerOutParameter()为输入输出参数注册为 JDBC 类型的 INTEGER。

```
cs.setInt(1, 1);
cs.registerOutParameter(1, Types.INTEGER);
```

(4) 执行存储过程。执行存储过程是调用 CallableStatement 对象的 execute()完成。例如以下语句使用 CallableStatement 对象 cs 完成存储过程的调用:

```
cs.execute();
```

(5) 获得输出参数值(可选)。如果调用的存储过程有输出参数，那么需要使用 CallableStatement 对象的 getter()方法获得输出参数值。

下面举例说明 JDBC 调用 MySQL 存储过程。在 MySQL 数据库 testDB 中，表 user 的 DDL 如下：

```
CREATE TABLE `user` (
  `id` int(11) NOT NULL AUTO_INCREMENT,
  `name` varchar(50) NOT NULL,
  `password` varchar(255) DEFAULT NULL,
  `sex` char(10) DEFAULT NULL,
  `age` int(11) DEFAULT NULL,
  `birthday` datetime DEFAULT NULL,
  PRIMARY KEY (`id`)
) ENGINE=InnoDB AUTO_INCREMENT=1 DEFAULT CHARSET=utf8;
```

存储过程 getUserNameAndAgeById_2 用于查询 user 表的 name 和 password 字段值，输出参数为 userName 和 userAge，输入参数为 userId。存储过程 getUserNameAndAgeById_2 代码如下：

```
CREATE DEFINER=`root`@`localhost` PROCEDURE `getUserNameAndAgeById_2`(
  IN `userId` int,OUT `userName` varchar(50),OUT `userAge` int)
BEGIN
  SELECT `name` INTO userName FROM `user` WHERE id =userId;
  SELECT age INTO userAge FROM `user` WHERE id =userId;
END
```

CallableStatementTest 类调用存储过程的方法 getUserNameAndAgeById_2()，其关键代码如下：

```
...
public class CallableStatementTest {
    public static void main(String[] args) {
        getUserNameAndAgeById_2();
    }
    public static void getUserNameAndAgeById_2() {
        Connection conn =null;
        CallableStatement cs =null;
        ResultSet rs =null;
        try {
        //存储过程函数格式:{call getUserNameAndAgeById_2(?,?,?)}
            conn =DBConnection.getConnection();
            //①定义调用存储过程的 SQL 语句
            String sql ="{call getUserNameAndAgeById_2(?,?,?)}";
            //②获取 CallableStatement 对象
```

```java
            cs = conn.prepareCall(sql);
            // ③设置参数(可选)
            cs.setInt(1, 1);
            cs.registerOutParameter(2, Types.VARCHAR);
            cs.registerOutParameter(3, Types.INTEGER);
            //④执行存储过程
            cs.execute();
            // ⑤获得输出参数值(可选)
            String userName = cs.getString(2);
            int userAge = cs.getInt("userAge");
            System.out.println(userName + "-" + userAge);
        } catch (SQLException e) {
            e.printStackTrace();
        } finally {
            DBConnection.close(rs, cs, conn);
        }
    }
}
```

2.5 JDBC事务管理

2.5.1 事务的概念

事务(Transaction)是一个包含一系列操作的不可分割的工作逻辑单元。这些操作作为一个整体一起向系统提交,要么都执行,要么都不执行。

事务具有ACID属性。ACID是原子性(atomicity)、一致性(consistency)、隔离性(isolation)、持续性(durability)的缩写。

(1) 原子性是指事务是一个完整的操作。事务的各步操作是不可分的(原子的);要么都执行,要么都不执行。

(2) 一致性是指当事务完成时,数据必须处于一致的状态。

(3) 隔离性是指对数据进行修改的所有并发事务是彼此隔离的,这表明事务必须是独立的,它不应以任何方式依赖于或影响其他事务。

(4) 持续性又称为持久性,指事务完成后对数据库的修改被永久保存。

2.5.2 JDBC事务管理

JDBC事务是指在数据库操作中,一项事务由一条或多条对数据库更新的SQL语句所组成的一个不可分割的工作单元;具有原子性,要么全部成功执行完毕;要么执行失败、撤销事务。

例如在银行转账时，张三把1000元转到李四账户上。使用SQL语句模拟如下：

```
update account set money=money-1000 where name=`张三`;
update account set money=money+1000 where name=`李四`;
```

这两条SQL语句组成了一项事务。或者两条SQL语句都成功执行，才能提交该事务；或者虽然执行了一条，整个事务必须全部撤销。否则，假设第一条执行成功，第二条执行失败：这时张三账号上少了1000元，但是李四账号上却没有增加1000元。

Java事务分为本地事务和全局事务两种类型。本地事务是指对单个数据源进行操作的事务。全局事务是指对多个数据源进行操作的事务。JDBC事务属于本地事务。Java程序如果需要全局事务应该使用JTA。

1. 设置事务提交方式

对于事务的原子性要求，java.sql.Connection类提供了如下控制事务的3个方法。

setAutoCommit(boolean autoCommit)：设置事务提交方式，该方法参数为true时表示事务是自动提交的。默认是自动提交事务。该方法参数为false时表示手动提交事务，需要提交事务时应调用commit()方法。

commit()：在设置了setAutoCommit(false)方法后，必须显式地调用该commit()方法提交事务。

rollback()：如果执行事务时出错，必须调用该方法进行数据库回滚，即将已经执行的事务中的部分操作撤销，使得数据库回到事务执行前的状态。

(1) 自动提交事务。前面使用JDBC访问数据库的例子都属于自动提交事务。如下代码属于自动提交事务：

```
public static void add() throws SQLException {
    Connection conn =null;
    PreparedStatement ps =null;
    ResultSet rs =null;
    try {
        conn =DBConnection.getConnection();
        //事务开始，默认自动提交事务
        //相当于使用了 conn.setAutoCommit(true);
        String sql ="insert into user"
                +"(id,name,password,sex,age,birthday)"
                +" values(?,?,?,?,?,?)";
        ps =conn.preparedStatement(sql);
        ps.setInt(1, 1);
        ps.setString(2, "zhangsan");
        …
        int lines =ps.executeUpdate();
        //每执行一条 SQL 语句事务自动提交
        //相当于使用了 conn.commit();
    } finally {
```

```
        DBConnection.close(rs, ps, conn);
    }
}
```

(2) 手动提交事务。如下代码模拟银行转账业务实现手动提交事务:

```
try {
    conn = DBConnection.getConnection();
    conn.setAutoCommit(false);//事务开始,设置手工提交事务

    String sql = "update user set money=money-? where id=?";
    ps = conn.prepareStatement(sql);
    ps.setFloat(1, m1);
    ps.setInt(2, id1);
    ps.executeUpdate();

    String sql2 = "update user set money=money+? where id=?";
    ps = conn.prepareStatement(sql2);
    ps.setFloat(1, m1);
    ps.setFloat(2, id2);
    ps.executeUpdate();
    conn.commit();
    //事务提交
    } catch(SQLException e){
    //事务中某项操作不成功事务回滚
        conn.rollback();
    }finally {
        DBConnection.close(rs, ps, conn);
    }
}
```

(3) 存储点。如果在事务管理时仅仅需要撤回到某个 SQL 执行点,那么可以设置存储点(Save Point)。例如,如下代码设置存储点后,如果事务失败将撤回到存储点。

```
SavePoint point = null;
try {
    conn.setAutoCommit(false);
    Statement stmt = conn.createStatement();
    stmt.executeUpdate("insert into …");
    …
    point = conn.savePoint();                    //设置存储点
    stmt.executeUpdate("insert  into …");
    …
    conn.commit();
    } catch(SQLException e) {
        e.printStackTrace();
        if(conn != null) {
```

```
            try {
                if(point ==null) {
                    conn.rollback();
                } else {
                    conn.rollback(point);           //撤回到存储点
                    conn.releaseSavePoint(point);   //释放存储点
                }
            }
        }
    }
```

2. 设置事务隔离级别

对于事务隔离性的要求，java. sql. Connection 接口的 getTransactionIslation()和 setTransactionIsolation()分别用于获取和设置隔离级别。Connection 接口定义的 5 个隔离级别说明见表 2-12。

表 2-12　Connection 接口定义的 5 个隔离级别说明

隔离级别	说　明
TRANSACTION_NONE	不支持事务
TRANSACTION_READ_UNCOMMITED	读未提交，指一个事务可以读取另一个未提交事务操作的数据。由此导致的脏读、不可重复读、幻读都是允许的
TRANSACTION_READ_COMMITED	读已提交，指一个事务仅仅允许读另一个事务已经提交的数据。不允许脏读，允许可重复读和幻读
TRANSACTION_REPEATABLE_READ	可重复读，指不允许脏读和不允许可重复读，允许存在幻读
TRANSACTION_ SERIALIZABLE	串行事务，指不允许发生脏读、不可重复读和幻读

2.6　JDBC 4. x

JDBC 4.0 随 JDK 1.6 发布。与 JDBC 3.0 相比，主要增加了 JDBC 驱动类的自动加载、连接管理的增强、对 RowId SQL 类型的支持、SQL 的 DataSet 实现使用了 Annotations、SQL 异常处理的增强、对 SQL XML 的支持。

JDBC 4.1 随 JDK 1.7 发布。与 JDBC 4.0 相比，主要更新了两个特性。

1. 使用 try 语句自动关闭资源对象

Connection、ResultSet 和 Statement 都实现了 Closeable 接口。在 try 语句中调用这些接口的实例时，会自动调用其 close()方法实现自动关闭相关资源。例如，如下代码用于判断能否获取数据库连接。

```
public class JDBC41Test {
    public static void main(String[] args) throws ClassNotFoundException,
            SQLException {
        //定义访问数据库相关参数如下
        String driverName="com.mysql.jdbc.Driver";
        String url ="jdbc:mysql://localhost:3306/testDB";
        String dbUser="root";
        String dbPassword="root";

        Class.forName(driverName);
        //获取数据库连接进行操作
        try(Connection conn =DriverManager.getConnection(url, dbUser,
            dbPassword)){
            System.out.printf("数据库已经%s%n",
                conn.isClosed() ?"关闭":"打开");
        }
    }
}
```

2. 创建 JDBC 驱动支持的各种 ROWSets

RowSet 1.1 引入 RowSetFactory 接口和 RowSetProvider 类,可以创建 JDBC 驱动支持的各种 ROWSets。例如如下代码使用默认的 RowSetFactory 实现创建一个 JdbcRowSet 对象。

```
RowSetFactory myRowSetFactory =null;
JdbcRowSet jdbcRs =null;
ResultSet rs =null;
Statement stmt =null;
try{
    myRowSetFactory =RowSetProvider.newFactory();//用默认的 RowSetFactory 实现
    jdbcRs =myRowSetFactory.createJdbcRowSet();

    //创建一个 JdbcRowSet 对象,配置数据库连接属性
    jdbcRs.setUrl("jdbc:myDriver:myAttribute");
    jdbcRs.setUsername(username);
    jdbcRs.setPassword(password);

    jdbcRs.setCommand("select ID from TEST");
    jdbcRs.execute();
}
```

小　结

本章介绍了 JDBC 概念以及 JDBC 3.0 API 中的主要类和接口，使用 JDBC 3.0 API 访问数据库从逻辑上通常需要的 5 个步骤。除了进行数据库表的增、删、改、查操作，JDBC 还提供了调用存储过程、使用预编译的 PreparedStatement 进行数据库操作。JDBC 事务管理是基于本地事务，不支持跨越多个数据库或资源的分布式事务管理。对于分布式事务管理需要使用 JTA。JDBC 4.0 对 JDBC 3.0 进行了进一步完善和增强。

思考与习题

1. 什么是 JDBC?

2. JDBC 驱动程序分为几种？现在常用的是哪一种？

3. 请写出使用 JDBC 3.0 访问数据库 5 个步骤的文字描述。

4. 根据本章 MySQL 数据库 testDB 中的表 user 设计，请写出 Java 代码实现使用 Statement 查询 user 表 id 为 1 的记录。

5. 根据本章 MySQL 数据库 testDB 中的表 user 设计，请写出 Java 代码实现使用 PreparedStatement 查询 user 表 id 为 1 的记录。

6. 编写存储过程 getAgeById，要求存储过程只有一个参数 inIdOutAge 且该参数为输入输出(INOUT)类型。作为输入参数时，inIdOutAge 表示用户 id，作为输出参数时 inIdOutAge 表示查询结果 age。

7. 使用 JDBC 完成对存储过程 getAgeById 的调用，要求将输出参数值打印到控制台。

8. 编写一个类 CRUDTestByPSAndDBConnection，使用数据库帮助类 DBConnection，完成使用 PreparedStatement 实现增、删、改、查。

9. 使用 JDBC 事务管理完成向数据库 user 表插入一条记录并更新该记录的 age 字段值。

10. 编写一个类 UseJDBC4，使用 JDBC 4.X 完成表 user 的增、删、改、查。

第3章

Servlet

Servlet 是 Java 服务器端的扩展技术。与传统的 CGI 技术相比，Servlet 具有效率高、容易使用、功能强大、可移植性好的优点。

3.1 Servlet 简介

3.1.1 Servlet 的概念

Servlet 是一种服务器端的 Java 程序，具有独立于平台和协议的特性，可以生成动态的 Web 页面。它充当浏览器或其他 HTTP 客户程序与服务器的中间层。与传统的从命令行启动的 Java 应用程序不同，Servlet 由 Web 服务器进行加载。该 Web 服务器必须安装支持 Servlet 的 Java 虚拟机。现在的大型应用程序中 Servlet 常常用作控制器。

Servlet 和客户的通信采用"请求/响应"模式。Servlet 运行在 Servlet 容器中，和 JSP 统称为 Java Web 层的两大组件。

3.1.2 Servlet 与 CGI 的区别

一个 Servlet 实例采用多线程的方式为多个客户服务，如图 3-1 所示。

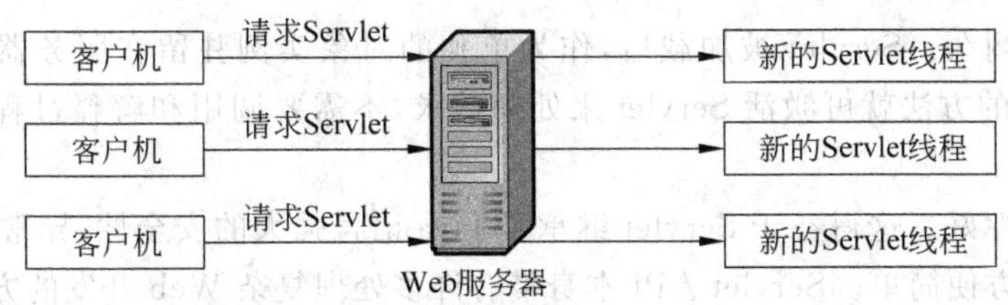

图 3-1　Servlet 采用多线程方式为多个客户服务

一个客户端请求一个 Servlet 时，Servlet 容器会创建该 Servlet 的一个实例并创建该 Servlet 实例的一个线程处理请求并返回响应。这和以前的 CGI 不同。在 Servlet 出现之前，实现 Web 应用动态功能需要使用 CGI。CGI 的一个重要缺点是对每一个客户请求在服务器端必须创建一个CGI进程。如果请求的客户很多,会导致服务器性能下降甚至崩溃,如图3-2所示。

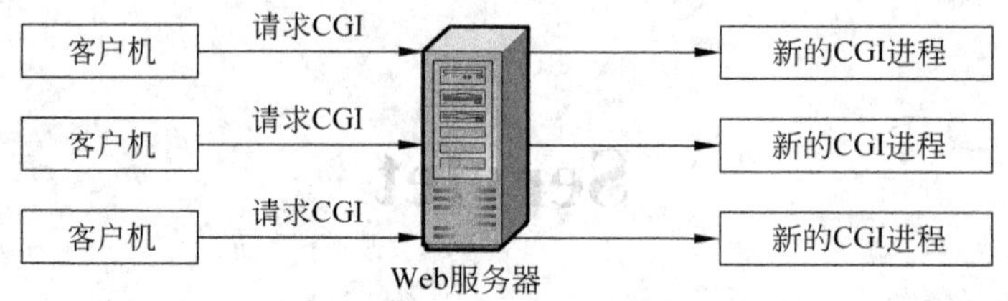

图3-2 CGI采用进程方式为客户服务

3.1.3 Servlet的功能

(1) 动态生成Web页面。Servlet能根据客户的请求,动态创建并返回一个HTML页面。这是Servlet最初的用法。

(2) 处理客户的HTML表单输入。Servlet能够接收客户HTML表单中的数据进行处理并进行适当的响应。

(3) 与服务器资源如数据库等交互。在Servlet中可以使用JDBC来访问数据库。

(4) 调用JavaBean或EJB组件执行业务逻辑。

(5) 视图派发功能。将JSP页面或HTML页面或Servlet等视图派发到客户端。

3.1.4 Servlet的优点

(1) 可移植性好。由于Servlet是由Java编写,因此Servlet可以跨平台。另外,Servlet在实现了Servlet规范的服务器上有很强的可移植性。例如,可以Windows平台上开发一个Servlet,将其放置到Windows的Tomcat上运行;也可以将其移植到UNIX系统的WebLogic服务器上运行。

(2) 功能强大。Java语言能实现的功能Servlet基本都能实现(除AWT和SWING图形界面外)。

(3) 高效耐久。Servlet被加载后,作为单独的对象实例驻留在服务器内存中,服务器只需要简单的方法就可激活Servlet来处理请求,不需要调用和解释过程,响应速度非常快。

(4) 安全性好。这得益于Servlet继承了Java语言强大的安全性、异常处理机制。

(5) 使用方便简单。Servlet API本身带有许多处理复杂Web开发的方法和类,例如获取HTML表单数据、读取HTTP头部、处理Cookie、会话跟踪等。

(6) 集成性好。Servlet由Servlet容器管理。Servlet能使用容器提供的功能(如数据源连接池等)。

3.1.5　Servlet API 简介

Servlet API 由 javax. servlet 和 javax. servlet. http 这两个包中的接口和类组成。javax. servlet 包中的接口和类定义了与具体协议无关的 Servlet;javax. servlet. http 包中的接口和类定义了采用 HTTP 协议进行通信的 Servlet。Servlet API 继承层次结构如图 3-3 所示。

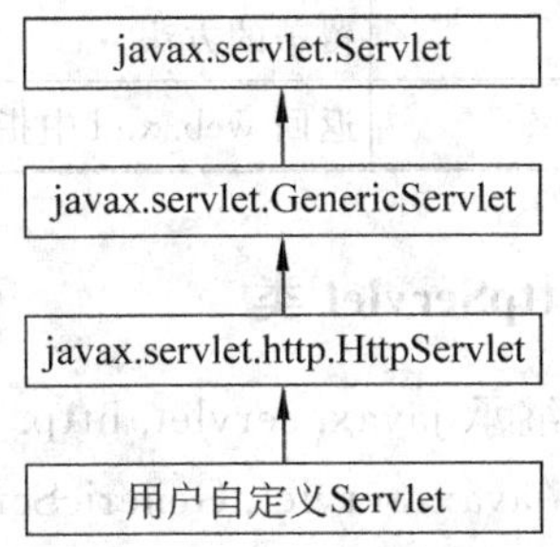

图 3-3　Servlet API 继承层次结构

1. javax. servlet. Servlet 接口

Servlet 规范要求所有的 Servlet 类必须直接或间接实现 javax. servlet. Servlet 接口，该接口的方法说明见表 3-1。

表 3-1　javax. servlet. Servlet 接口的方法说明

方　　法	方法说明
init()	进行初始化操作
service()	响应客户请求为客户服务
destroy()	释放 Servlet 占用的资源
getServletConfig()	返回 ServletConfig 对象,该对象代表了 Servlet 的配置信息
getServletInfo()	获取 Servlet 的文本信息,如作者、版本、版权等信息

在实现该接口时,必须实现该接口的 init()方法、service()方法、destroy()方法。通过直接实现 Servlet 接口来创建 Servlet 的方式实际中很少用到。

2. javax. servlet. GenericServlet 类

javax. servlet. GenericServlet 类实现了 javax. servlet. Servlet 接口。如果需要开发与具体协议无关的 Servlet,那么就可以通过继承 GenericServlet 并覆盖 GenericServlet 的 service(ServletRequest request, ServletResponse response)方法。通过继承 GenericServlet 来创建 Servlet 的方式实际中也很少用到。

javax. servlet. GenericServlet 类的常用方法说明见表 3-2。

表 3-2 javax. servlet. GenericServlet 类的常用方法说明

方　　法	方法说明
String getInitParameter(String name)	返回参数 name 指定的初始化参数值
ServletConfig getServletConfig()	返回 ServletConfig 对象，该对象代表 Servlet 配置对象
ServletContext getServletContext()	返回代表 Servlet 上下文的 ServletContext 对象
abstract void service(ServletRequest request, ServletResponse response)	Servlet 容器调用该方法响应客户请求；该方法是 GenericServlet 的唯一抽象方法，也是必须要被子类所覆盖的方法
String getServletName()	返回 web. xml 中指定的 Servlet 的名字

3. javax. servlet. http. HttpServlet 类

大多数情况下，开发者需要继承 javax. servlet. http. HttpServlet 来创建基于 HTTP 协议的 Servlet。HttpServlet 是 javax. servlet. GenericServlet 的子类，重写了从父类继承的方法（如 service()方法等），并增加了新的方法（如 doGet()、doPost()、doPut()、doDelete()方法等）。HttpServlet 类的常用方法说明见表 3-3。

表 3-3 javax. servlet. http. HttpServlet 类的常用方法说明

方　　法	方法说明
void doGet(HttpServletRequest request, HttpServletResponse response)	在接收 HTTP get 请求时，Servlet 容器调用该方法处理 get 请求
void doPost(HttpServletRequest request, HttpServletResponse response)	在接收 HTTP post 请求时，Servlet 容器调用该方法处理 post 请求

继承 HttpServlet 类来派生 HttpServlet 子类时，如果处理的是 HTTP get 请求，那么就在子类中重写 doGet()方法；如果处理的是 HTTP 其他请求，那么就在子类中重写其他对应的方法。

4. HttpServletRequest 和 HttpServletResponse 接口

在以上 HttpServlet 的方法中，HttpServletRequest request 对象包含了客户端的请求信息，HttpServletResponse response 对象包含了返回到客户端的响应信息。在接收到客户端请求时，Servlet 容器会创建 HttpServletRequest 的实例并将请求信息封装到其属性中，创建 HttpServletResponse 的实例并将返回到客户端的响应信息封装到其属性中。

在 Servlet API 中，HttpServletRequest 和 HttpServletResponse 是接口。它们分别继承自 javax. servlet. ServletRequest 和 javax. servlet. ServletResponse 接口。javax. servlet. ServletRequest 和 javax. servlet. ServletResponse 接口的实例代表与具体协议无关的请求和响应对象；HttpServletRequest 和 HttpServletResponse 代表基于 HTTP 协议通信的请求和响应对象。

javax. servlet. ServletRequest 接口的常用方法说明见表 3-4。

表 3-4　javax. servlet. ServletRequest 接口的常用方法说明

方　法	方法说明
void setCharacterEncoding(java. lang. String env)	设置请求的字符编码为参数指定的 env
void setAttribute(String name, Object o)	将参数指定的 o 对象保存在 ServletRequest 对象中，同时指定参数 name 为保存的对象的名字
Object getAttribute(String name)	根据参数 name，返回在 ServletRequest 对象中保存的对象
String getParameter(String name)	返回请求中参数 name 值
Enumeration getParameterNames()	返回表示请求中的所有参数的名字的一个 Enumeration 对象
String[]getParameterValues(String name)	返回表示请求中 name 参数的所有值的 String 数组对象

javax. servlet. ServletResponse 接口的常用方法说明见表 3-5。

表 3-5　javax. servlet. ServletResponse 接口的常用方法说明

方　法	方法说明
String getContentType()	返回响应的 MIME 类型
ServletOutputStream getOutputStream()	返回向客户端发送二进制数据的输出流 ServletOutputStream
PrintWriter getWriter()	返回向客户端发送字符数据的 PrintWriter 对象

javax. servlet. http. HttpServletRequest 接口的常用方法说明见表 3-6。

表 3-6　javax. servlet. http. HttpServletRequest 接口的常用方法说明

方　法	方法说明
String getMethod()	返回 HTTP 请求方法(如 get、post 等)
HttpSession getSession()	返回当前请求对象的 HttpSession 对象
String getServletPath()	返回请求 URL 上下文中的子串，不包括 Servlet 上下文(即 Web 应用)和查询字符串
String getRequestURI()	返回请求 URL 的一部分，包括上下文但是不包括任意查询字符串
String getQueryString()	返回请求 URL 中的查询字符串
String getContextPath()	返回 Servlet 上下文(即 Web 应用)的 URL 的前缀
Cookies[] getCookies()	返回一个与请求相关的 Cookies 数组
String getHeader(String name)	返回参数 name 指定的 HTTP 报头名称
Enumeration getHeaders(String name)	返回请求中 name 指定的 HTTP 报头的所有枚举值
Enumeration getHeaderNames()	返回请求中所有 HTTP 报头名称的枚举值

javax. servlet. http. HttpServletResponse 接口的常用方法说明见表 3-7。

表 3-7　javax.servlet.http.HttpServletResponse 接口的常用方法说明

方　法	方法说明
void addCookie(Cookie cookie)	将参数 cookie 添加到响应中
void setHeader(String name,String value)	将响应中报头名称 name 的值设置为 value

3.1.6　Servlet 的生命周期

1. Servlet 生命周期的 3 个阶段

Servlet 部署在 Servlet 容器中，由 Servlet 容器管理。Servlet 的生命周期指 Servlet 容器加载 Servlet、初始化 Servlet、Servlet 接收请求、Servlet 提供服务、Servlet 容器销毁的整个过程。整个过程分为如下 3 个阶段。

(1) 初始化阶段。Servlet 容器负责加载 Servlet 类文件、创建 Servlet 类的实例、调用 Servlet 实例的 init()方法。

(2) 响应客户请求阶段。Servlet 容器收到客户请求后，生成请求对象 ServletRequest，将请求对象 ServletRequest 传递给 Servlet 实例的 service()方法。service()方法负责从请求对象 ServletRequest 中获得请求信息并处理请求。

(3) 终止阶段。当 Web 应用被终止或 Servlet 容器停止运行或 Servlet 容器重新装载该 Servlet 时，Servlet 容器会调用 Servlet 的 destroy()方法释放 Servlet 所占的资源。

Servlet 的生命周期如图 3-4 所示。

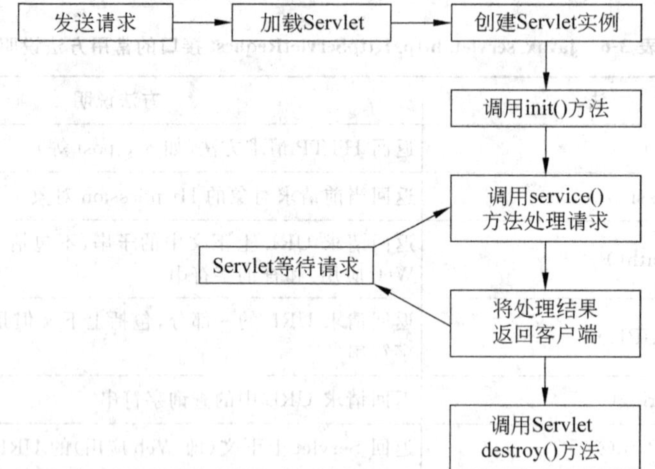

图 3-4　Servlet 的生命周期

2. Servlet 容器加载 Servlet 的 3 种情形

Servlet 容器启动时自动加载 Servlet。这需要在 web.xml 文件中配置该 Servlet 的<load-on-startup>属性设置。

Servlet 容器启动后客户首次请求未加载的 Servlet 时，Servlet 容器会加载该 Servlet。

Servlet 的类文件被更改后 Servlet 容器重新加载更新的版本。这通常在程序开发和调试阶段进行。

3. Web 服务器的两种启动模式

（1）调试模式。用于应用开发和调试阶段。在调试模式下，Web 服务器会自动扫描 Servlet 类文件是否修改。如果有 Servlet 类文件修改，会加载新版本的 Servlet。自动扫描和加载 Servlet 会影响服务器性能。

（2）运行模式。用于应用正常运行阶段，确保 Servlet 类文件修改后不再自动扫描和加载。

4. Tomcat 设置 Web 应用的启动模式

在 Web 服务器 Tomcat 的%Tomcat%conf/server.xml 文件中，<Context>节点代表某个 Web 应用。<Context>节点的 reloadable 属性设置为 true，表示设置该 Web 应用的启动模式为调试模式，即 Tomcat 会自动扫描和加载该 Web 应用中修改的 Servlet；<Context>节点的 reloadable 属性设置为 false，表示设置该 Web 应用的启动模式为运行模式，即 Tomcat 不会自动扫描和加载该 Web 应用中修改的 Servlet。

3.2 创建 Servlet

3.2.1 Java Web 应用的目录结构

1. Java Web 应用的文件组成

在一个 Java Web 应用中，除了 Servlet、JSP 等 Web 组件之外，还包括 Web 层的部署描述文件 web.xml(可选，自 Java 5)、应用需要的其他辅助类(如数据库连接类、在 JSP 页面和 Servlet 之间传输数据及在 Servlet 和数据访问层之间传输数据的 JavaBean)、静态资源文件 HTML 页面和图片文件等。这些文件存放位置必须符合 Java Web 应用的目录结构要求。

2. Java Web 应用的目录结构

Java Web 应用的目录结构如图 3-5 所示。

（1）Web 应用根目录。在该目录下有 WEB-INF 目录；JSP 文件、静态 HTML 文件及 Applet 类文件需要放在此目录下，或在此目录下再创建分类目录分别存放相关文件。例如，创建 jsp 目录存放 JSP 文件、创建 images 目录存放图片文件、创建 css 目录存放 CSS 文件。

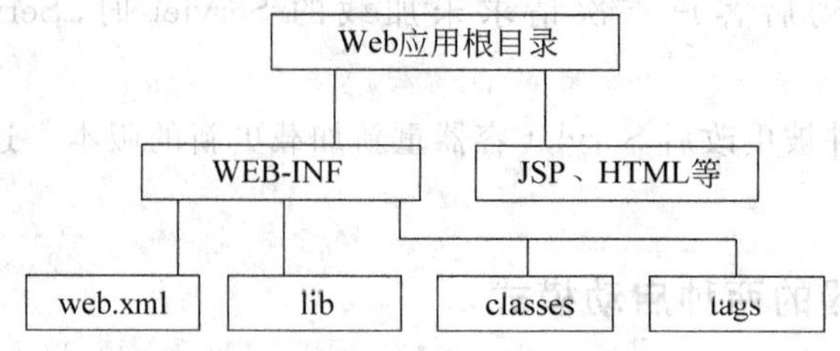

图 3-5　Java Web 应用的目录结构

(2) WEB-INF 目录。Web 层的部署描述文件 web.xml 放在此目录下。

(3) lib 目录。存放该 Java Web 应用需要的第三方类库,类库通常是 jar 文件。

(4) classes 目录。存放该 Java Web 应用编译后的 Java 类文件。

(5) tags 目录。存放该 Java Web 应用的标签库描述文件。

3. Tomcat Web 服务器自带的应用

对于 Tomcat Web 服务器,Web 应用通常部署到%Tomcat%\webapps\。webapps 目录下的每个子目录都是一个 Java Web 应用程序。Tomcat 自带了多个帮助学习 Java Web 知识和管理 Tomcat 的应用,如图 3-6 所示。

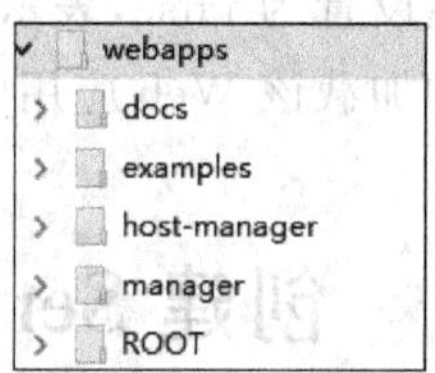

图 3-6　Tomcat 自带的 Web 应用

3.2.2　创建和配置 Servlet

通常自定义的 Servlet 在创建后需要配置,便于使用路径访问 Servlet。下面将编写第一个 Servlet——HelloworldServlet,用于向浏览器返回一个显示 HelloWorld 的 HTML 页面。

1. 创建 Servlet

HelloWorldServlet 的关键代码如下。

```
package com.test.servlet.HelloWorldServlet;
...
public class HelloWorldServlet extends HttpServlet{
    public void doGet(HttpServletRequest request, HttpServletResponse response)
        throws ServletException,IOException{
        doPost(request,response);
```

```
    }
    public void doPost(HttpServletRequest request, HttpServletResponse response)
        throws ServletException,IOException{
        //设置响应类型
        response.setContentType("text/html");
        //通过 response 对象获得 PrintWriter 的对象 out
        PrintWriter out=response.getWriter();
        //使用 out 输出一个 HTML 页面
        out.println("<html><head><title>HelloWorld</title></head>"+
            "<body><h1>HelloWorld!</h1></body></html>");
    }
}
```

HelloWorldServlet 继承了 HTTP Servlet，用于处理 Web 应用客户端和服务器之间使用 HTTP 协议的客户请求。

在 HelloWorldServlet 中，doGet()方法调用 doPost()方法。

doPost()方法中使用输出流对象 PrintWriter out 向客户端返回一个显示“HelloWorld!”的 HTML 页面。

2. 配置 Servlet

配置 Servlet 有两种方式。

(1) 使用 XML 配置 Servlet。在 Servlet 2.5 规范之前(包括 2.5)，需要在 Java Web 层的部署描述文件 web.xml 中进行配置。Tomcat 6 是支持 Servlet 2.5 规范的 Java Web 服务器。

在 web.xml 中 HelloWorldServlet 配置如下：

```
<servlet>
    <servlet-name>hello</servlet-name>
    <servlet-class>com.test.servlet.HelloWorldServlet</servlet-class>
</servlet>
<servlet-mapping>
    <servlet-name>hello</servlet-name>
    <url-pattern>/helloworld</url-pattern>
</servlet-mapping>
```

<servlet>元素用于声明一个 Servlet。它包括两个必需的子元素：<servlet-name>子元素声明该 Servlet 的名字；<servlet-class>子元素指定该 Servlet 的全限定名。

<servlet-mapping>元素用于配置一个 Servlet 的映射路径。它包括两个必需的子元素：<servlet-name>子元素表示需要配置映射路径的 Servlet 的名字；<url-pattern>子元素指定该 Servlet 的映射路径。

(2) 使用注解配置 Servlet。在 Servlet 3.0 规范后(包括 3.0)可以使用注解配置 Servlet，也可以使用 XML 配置 Servlet。使用注解配置 HelloWorldServlet 如下：

```
package com.test.servlet.HelloWorldServlet;
...
@WebServlet(name ="hello", urlPatterns ={ "/hello"})
public class HelloWorldServlet extends HttpServlet{
    ...
}
```

另外还可以使用 urlPatterns 配置 Servlet 的映射路径为多个。如下代码使用 urlPatterns 将 HelloWorldServlet 的映射路径配置为/hello 和/：

```
@WebServlet(name ="hello", urlPatterns ={ "/hello", "/" })
public class HelloWorldServlet extends HttpServlet{
    ...
}
```

这样通过 http://localhost:8080/ch03-1/hello 或 http://localhost:8080/ch03-1/均可访问 HelloWorldServlet。

(3) 配置 Servlet 的选择。对于使用 Servlet 3.0 规范以上的 Servlet 容器，建议使用注解配置 Servlet。这种方式简单方便。对于需要在 Servlet 2.5 容器中运行的应用或者没有源码的 Servlet，就必须使用 XML 配置 Servlet。

3.2.3 使用 Eclipse 创建和配置 Servlet

实际开发中通常使用集成开发环境 Eclipse 创建、部署和测试包含 Servlet 的 Java Web 项目，这样可以提高开发效率。

1. 创建 Java Web Project

在 Eclipse 菜单栏单击 File→New→Project→Web→Dynamic Web Project，单击 Next 按钮后，在弹出的 New Dynamic Web Project 窗口输入项目名称 ch03-01，单击 Finish 按钮完成项目创建工作。

2. 创建并配置 HelloWorldServlet

1) 打开创建 Servlet 向导

右击 ch03-01 项目中的 Java Resources\src，选择 New→Servlet 命令，打开创建 Servlet 向导，如图 3-7 所示。

2) 配置 Servlet 包名和类名

在创建 Servlet 向导打开的创建 Servlet 窗口中，配置包名和类名，如图 3-8 所示。在 Java package 文本框输入包名 servlet，在 Class name 文本框输入类名 HelloWorldServlet。

单击 Next 按钮进入配置 Servlet 映射信息窗口。

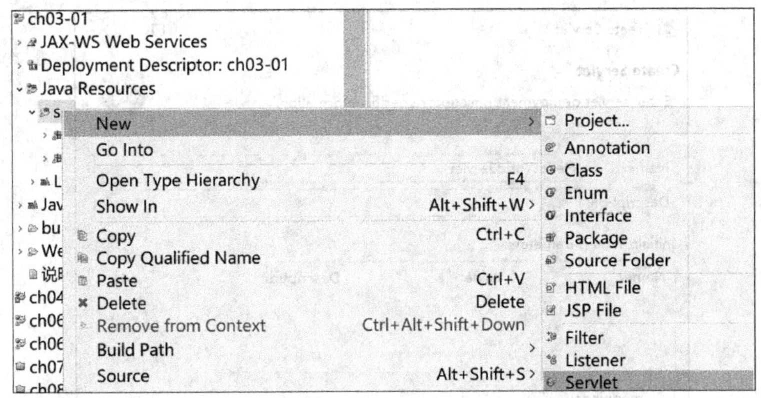

图 3-7 打开 Servlet 创建向导

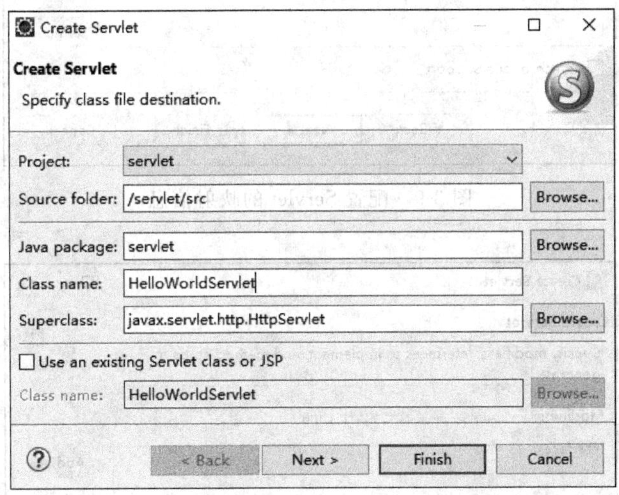

图 3-8 配置 Servlet 包名和类名

3）配置 Servlet 映射信息

配置 Servlet 映射信息包括配置名字和映射路径。Eclipse 自动生成的 Servlet 映射信息如图 3-9 所示。这些信息指定了 Servlet 的名字是 HelloWorldServlet，映射路径为/HelloWorldServlet。可以在选中 URL mappings 下方的/HelloWorldServlet 后单击右方的 Edit 按钮，修改 HelloWorldServlet 的映射路径为/hello。

单击图 3-9 中的 Next 按钮进入配置 Servlet 修饰符和选择 Servlet 方法的窗口，如图 3-10 所示。

在图 3-10 中选择 Inherited abstract methods、doGet 方法即可，其他选项在此不需要选中。最后单击 Finish 按钮，Eclipse 会自动打开 HelloWorldServlet. java 的编辑窗口。这时项目和编辑窗口出现项目出错的红色错误提示，这是没有配置构建路径中需要的 Servlet API 库导致的。需要在项目构建路径配置 Tomcat 的 servlet-api 库。

HelloWorldServlet. java 文件是 Servlet 类的源码，其关键代码如下：

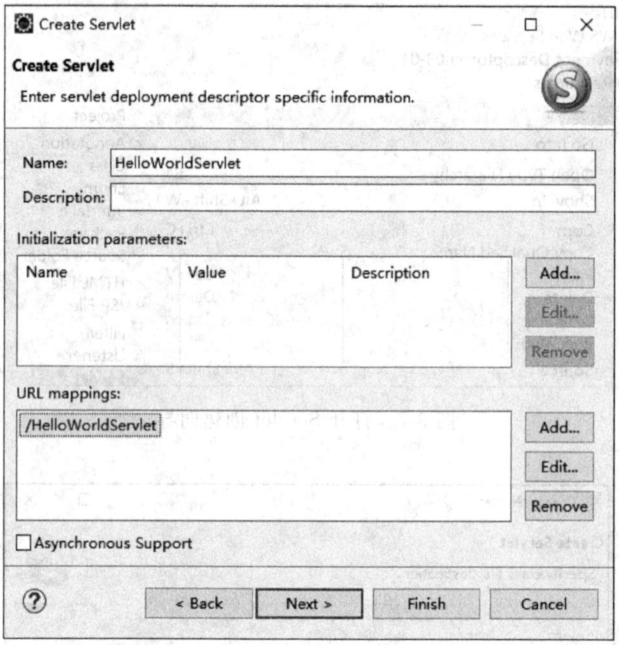

图 3-9　配置 Servlet 的映射信息

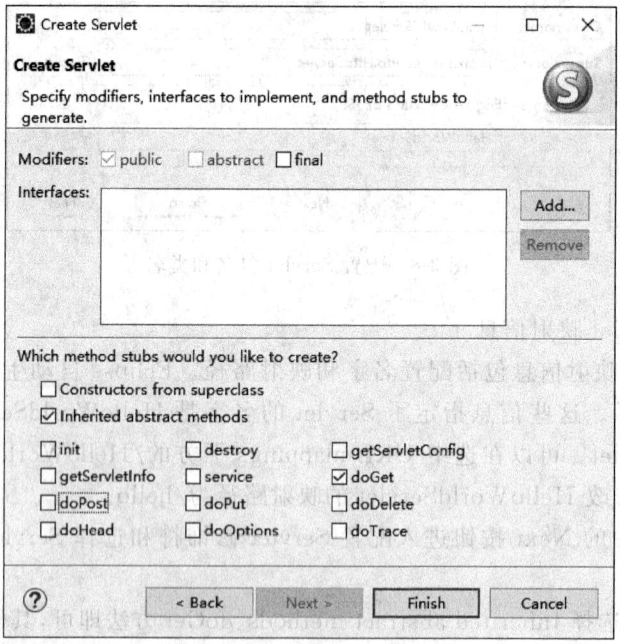

图 3-10　配置 Servlet 修饰符和方法

```
package servlet;
...
@WebServlet(name ="hello", urlPatterns ={ "/hello"})
```

```
public class HelloWorldServlet extends HttpServlet {
    public void doGet(HttpServletRequest request, HttpServletResponse response)
        throws ServletException, IOException {
        response.setContentType("text/html");
        PrintWriter out = response.getWriter();
        out.println("<html><head><title>A Servlet</title></head><body>");
        out.print("HelloWorld!");
        out.println("</body></html>");
        out.flush();
        out.close();
    }
}
```

3. 将项目部署到 Tomcat 下并运行

右击 servlet 项目，选择 Run As→Run on Server 命令后，Eclipse 会自动部署项目到 Tomcat，启动 Tomcat 运行该项目。

4. 打开浏览器进行测试

在浏览器地址栏中输入 http://localhost:8080/ch03-1/hello 后按 Enter 键，HelloWorldServlet 在浏览器中的显示效果如图 3-11 所示。

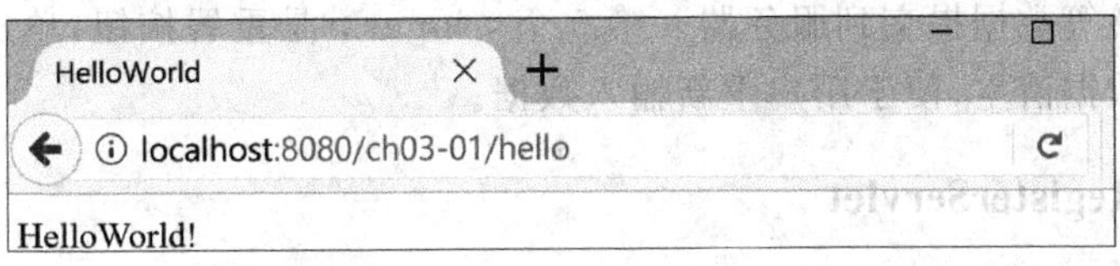

图 3-11 HelloWorldServlet 在浏览器中的显示效果

3.3 Servlet 常用功能

3.3.1 Servlet 接收请求参数

客户端提交数据可以通过 Servlet 的 getParameter(String paramName)方法得到。此处以用户注册为例讲解 Servlet 接收请求参数。

1. 创建和编辑注册页面

(1) 新建文件夹。在 3.2.3 节创建的 ch03-01 项目下的 WebRoot 下新建文件夹 register。右击 WebRoot→New→Folder 命令，弹出 New Folder 窗口，在其下方的 Folder name 文本框中输入 register，单击下方的 Finish 按钮。

(2) 新建页面 register.html。在文件夹 register 下新建 register.html。右击文件夹

register，依次选择 New→Web→HTML File 命令，单击 Next 按钮，打开创建 HTML 页面向导。在打开窗口中的 File name 输入框输入文件名 register，单击 Finish 按钮结束 register.html 的创建。对 register.html 进行编辑后，其关键代码如下：

```
<html>
    ...
<body>
    <center>
        <form action="Register" method="post">
            用户注册<br>
            姓名:<input type="text" name="name"><br>
            密码:<input type="password" name="password"><br>
            年龄:<input type="text" name="age"><br>
                <input type="submit" value="注册">
                    <input type="reset" value="重置">
            </form>
        </center>
    </body>
</html>
```

register.html 页面中的 Form 表单用于向服务器提交数据。Form 表单的前 3 个输入框分别用于输入用户名、密码、年龄。第 4 个<input>是注册按钮，单击该按钮会将用户名、密码、年龄等数据提交到服务器。第 5 个<input>是重置按钮，单击该按钮会将用户输入的所有数据清空，便于用户重新输入数据。

2. 创建 RegisterServlet

RegisterServlet 用于接收用户注册信息并将注册信息打印到控制台，其关键代码如下：

```
package servlet;
...
@WebServlet("/register/Register")
public class RegisterServlet extends HttpServlet {
    public void doPost (HttpServletRequest request, HttpServletResponse
response)
        throws ServletException, IOException {
        //设置接收请求信息的编码格式
        request.setCharacterEncoding("UTF-8");
        //接收 register.html 的注册信息
        String name =request.getParameter("name");
        String password =request.getParameter("password");
        String sage =request.getParameter("age");
        //在控制台打印输出注册信息
        System.out.println("name=" +name +";password=" +password +
            ";sage=" +sage);
```

```
    }
}
```

@WebServlet("/register/Register")将 RegisterServlet 映射到路径/register/Register。

服务器将接收到的数据封装到 HttpServletRequest 对象 request 中,向客户端返回的响应信息封装到 HttpServletResponse 对象 response 中,并将这两个对象传递给 RegisterServlet 的 doPost()方法。

```
String name =request.getParameter("name");
String password =request.getParameter("password");
String sage =request.getParameter("age");
```

以上代码使用 Servlet 的 getParameter(String paramName)方法分别得到请求参数 name、password、age 的值。

3. 部署项目到 Tomcat

选择 Web 项目的 servlet,右击该项目,选择 Run As→Run on Server 命令,选择 Tomcat 后单击 Finish 按钮,Eclipse 会将该项目部署到 Tomcat,启动 Tomcat 并运行该应用。

4. 打开浏览器输入地址进行测试

在浏览器地址栏输入 http://localhost:8080/ch03-1/register/register.html 后按 Enter 键。如果程序正常运行,Eclipse 控制台会打印注册页面 register.html 输入的注册信息。

3.3.2 作用域与存取数据

1. Java Web 的 4 个作用域

在 Java Web 应用中,数据可以保存在 4 个作用域中:页面(page)作用域、请求(request)作用域、会话(session)作用域、应用(application)作用域。一般情况下,这 4 个作用域按从小到大顺序依次是页面作用域、请求作用域、会话作用域、应用作用域。JSP 中可以使用这 4 个作用域。Servlet 中能用的作用域是后 3 个。

HttpServletRequest、HttpSession、ServletContext 接口定义的 setAttribute()方法和 getAttribute()方法用于在请求作用域、会话作用域、应用作用域保存和获取数据。下面给出了使用这 3 个接口的实例来存取 User 对象的示例代码。

2. 在请求作用域中保存和获取对象

```
public class ScopeServlet extends HttpServlet {
    public void doPost(HttpServletRequest request, HttpServletResponse response)
```

```
        throws ServletException {
        ...
        User user =new User();
        user.setName("Mike");
        user.setPassword("123");
        user.setAge(22);
        request.setAttribute("user", user);
        ...
        User u = (User) request.getAttribute("user");
        ...
    }
}
```

以上代码使用 HttpServletRequest request 对象的 setAttribute("user", user)将 user 对象保存到请求作用域,且保存时对象的名字是 String 类型的 user。最后使用 request. getAttribute("user")来根据名字 user 获取保存在请求作用域中的 User 对象。request. getAttribute("user")返回的是 Object 类型,故需要转型为 User。

3. 在会话作用域中保存和获取对象

```
public class ScopeServlet extends HttpServlet {
    public void doPost(HttpServletRequest request, HttpServletResponse response)
        throws ServletException {
        ...
        User user =new User();
        user.setName("Mike");
        user.setPassword("123");
        user.setAge(22);
        HttpSession session =request.getSession();
        session.setAttribute("user", user);
        ...
        User u = (User) session.getAttribute("user");
        ...
    }
}
```

以上代码首先使用 HttpServletRequest. getSession()方法获取 HttpSession 的 session 对象,再使用 session 对象的 setAttribute("user", user)将 user 对象保存到会话作用域,且保存时对象的名字是 String 的 user。最后使用 session. getAttribute("user")来根据名字 user 获取保存在会话作用域中的 user 对象。session. getAttribute("user")返回的是 Object 类型,故需要转型为 User。

4. 在应用作用域中保存和获取对象

```
public class ScopeServlet extends HttpServlet {
```

```
public void doPost(HttpServletRequest request, HttpServletResponse response)
        throws ServletException {
        ...
        User user = new User();
        user.setName("Mike");
        user.setPassword("123");
        user.setAge(22);
        ServletContext application = this.getServletContext();
        application.setAttribute("user", user);
        ...
        User u = (User) application.getAttribute("user");
        ...
    }
}
```

以上代码首先使用 HttpServlet. getServletContext()方法获取 ServletContext 类型的对象 application,再使用 application 对象的 setAttribute("user", user)将 user 对象保存到应用作用域,且保存时对象的名字是 String 类型的 user。最后使用 application. getAttribute("user")来根据名字 user 获取保存在应用作用域中的 user 对象。application. getAttribute("user")返回的是 Object 类型,故需要转型为 User。

3.3.3 Servlet 请求转发与重定向

Servlet 可以将请求转发或重定向到另一个资源。

1. 请求转发

在 Servlet 中,使用

```
ServletRequest.getRequestDispatcher("目标地址").forward(request, response)
```

将请求转发到目标地址所代表的资源。Servlet 请求转发在服务器端执行,不会丢失请求信息。

例如,如下代码将请求从当前 Servlet 转发到另一个页面 registerSuccess. jsp。

```
public class RegisterServlet extends HttpServlet {
    public void doPost(HttpServletRequest request,
        HttpServletResponse response) throws ServletException {
        ...
        request.getRequestDispatcher("registerSuccess.jsp").forward(request,
response);
    }
}
```

在 doPost 方法中,request 对象的 getRequestDispatcher("registerSuccess. jsp")方法

获取一个 RequestDispatcher 对象，目标地址是 registerSuccess. jsp；接着调用 RequestDispatcher 对象的 forware（request，response）方法将请求转发到 registerSuccess. jsp。由于转发时携带着 request 和 response 对象，因此这种方式不会丢失请求信息。

2. 重定向

在 Servlet 中，使用 response. sendRedirect("目标地址")重定向到目标地址代表的资源。重定向会丢失原来的请求信息。

例如，如下代码从当前 Servlet 重定向到 registerSuccess. jsp。

```
public class RegisterServlet extends HttpServlet {
    public void doPost(HttpServletRequest request,
        HttpServletResponse response) throws ServletException {
        …
        response.sendRedirect("registerSuccess.jsp");
    }
}
```

3.3.4 获取 Servlet 初始化参数

使用 XML 配置或者使用注解配置 Servlet 时，可以配置 Servlet 运行时需要的参数。以下代码使用 XML 配置 Servlet，在 web. xml 中设置了 DBConnectingServlet 的初始化参数。

```
<servlet>
    <servlet-name>DBConnectingServlet</servlet-name>
        <servlet-class>util.DBConnectionServlet</servlet-class>
    <init-param>
        <param-name>driverName</param-name>
        <param-value>com.mysql.jdbc.Driver</param-value>
    </init-param>
</servlet>
```

子元素<init-param>用于设置 Servlet 的初始化参数。

<param-name>元素设置参数名为 driverName。

<param-value>元素设置参数值为 com. mysql. jdbc. Driver。

DBConnectingServlet 读取初始化参数的关键代码如下：

```
public class DBConnectingServlet extends HttpServlet {
    public void doPost(HttpServletRequest request, HttpServletResponse response)
        throws Exception {
        …
        this.getInitParameter("driverName");
```

```
    }
}
```

通过 DBConnectingServlet 对象的 getInitParameter（"driverName"）方法，获得了 web. xml 中设置的初始化参数值 com. mysql. jdbc. Driver。

3.3.5 配置 Servlet 加载顺序

如果在某个 Web 应用中存在多个 Servlet，且这多个 Servlet 之间有依赖关系，那么可以使用 XML 或注解配置 Servlet 容器加载 Servlet 的顺序。

1. 使用 XML 配置 Servlet 加载顺序

在 web. xml 中，使用 Servlet 的<load-on-startup>属性来配置 Servlet 容器加载 Servlet 的顺序。

例如，现在某个 Web 应用中存在两个 Servlet：FirstServlet 和 SecondServlet。SecondServlet 需要使用 FirstServlet 中的资源，只有 FirstServlet 首先加载，SecondServlet 才能正常加载，否则会出错。这两个 Servlet 在 web. xml 中配置如下：

```
<servlet>
   <servlet-name>FirstServlet</servlet-name>
   <servlet-class>servlet.FirstServlet </servlet-class>
   <load-on-startup>1</load-on-startup>
</servlet>
<servlet>
   <servlet-name>SecondServlet</servlet-name>
   <servlet-class>servlet.SecondServlet</servlet-class>
   <load-on-startup>2</load-on-startup>
</servlet>
```

FirstServlet 的<load-on-startup>属性设置为 1，SecondServlet 的<load-on-startup>属性设置为 2。属性值小的先加载。这样就保证了 FirstServlet 加载后再加载 SecondServlet。

2. 使用注解配置 Servlet 加载顺序

如果使用注解配置 Servlet 的加载顺序，则需要使用@WebServlet 的 loadOnStartup 属性。

例如，配置 FirstServlet 和 SecondServlet 启动顺序的关键代码如下：

```
@WebServlet(name ="FirstServlet", urlPatterns ={ "/firstServlet" },
loadOnStartup =1)
public class FirsteServlet extends HttpServlet {
   …
}
```

```java
@WebServlet(name ="SecondServlet", urlPatterns ={ "/secondServlet" },
loadOnStartup =2)
public class SecondServlet extends HttpServlet {
    ...
}
```

小　　结

作为 Web 层组件的 Servlet 现在主要用作控制器，实际开发中常常通过继承 HttpServlet 来创建。Servlet 3.0 规范中配置一个 Servlet 可以使用注解或 XML。Servlet 编程经常用到的功能包括接收请求参数、存取属性对象、视图派发、获取初始化参数等。为了熟练使用 Eclipse 进行 Servlet 编程，需要进行大量编程练习。

思考与习题

1. 什么是 Servlet?
2. 在为多个客户服务时，Servlet 与 CGI 有何区别?
3. 画图并说明 Servlet API 继承层次结构。
4. 通过实现 javax. servlet. Servlet 接口来创建 Servlet 时需要实现哪 3 个方法?
5. 什么是 Servlet 的生命周期?
6. 画图并说明 Java Web 应用的目录结构。
7. 配置 Servlet 有几种方式? 请编写代码分别举例说明。
8. 写出使用 Eclipse 开发一个 Servlet 的步骤。
9. 编写一个用户可以输入姓名、年龄、出生日期的注册页面，并编写接收注册数据的 RegisterServlet，实现将注册数据插入数据库 testDB 的 user 表。
10. 举例说明 ServletContext、HttpSession、HttpServletRequest 的 setAttribute()方法和 getAttribute()方法的使用。
11. 使用 Servlet 进行服务器端的视图派发时需要使用 Servlet API 的哪个方法?
12. 使用 Servlet 进行客户端的视图派发时需要使用 Servlet API 的哪个方法?
13. 举例说明如何获取配置 Servlet 时的初始化参数。
14. 举例说明有多个 Servlet 时如何配置加载顺序。

第4章 JSP

JSP 是 Sun 公司推出的一种动态网页技术标准。自 JSP 诞生以来，在电子商务等动态网站上得到了广泛应用。现在 JSP 还是 Java Web 开发中最常用的技术之一。

4.1 JSP 简介

4.1.1 JSP 的概念

JSP(Java Server Pages)是 Sun 公司制定的 Java Web 层组件标准，适合用作表示组件，主要用来创建动态网页如网上购物等应用。JSP 建立在 Servlet 技术之上，与 Servlet 相比，JSP 更适合作为表示组件。

JSP 文件由 JSP 元素和模板元素构成。JSP 元素包括注释、指令、脚本元素、动作元素、表达式语言、标准标签库，由 JSP 引擎直接处理。模板元素可以是 HTML 元素或是 XML 元素。例如，如下是一个 JSP 页面 HelloWorld.jsp 的关键代码：

```
<%@page language="java" contentType="text/HTML; charset=UTF-8"
    pageEncoding="UTF-8"%>
<html>
…
<body>HelloWorld!</body>
</html>
```

4.1.2 JSP 的优点

(1) 内容的生成和显示分离。美工使用 HTML 或者 XML 标记来设计和美化页面，而程序员使用 JSP 元素来生成页面上的动态内容。

(2) 强调可重用的组件。除了强调 JSP 本身的可重用性，还可以通过与可重用的、跨平台的业务逻辑组件 JavaBean 或 EJB 结合，完成业务逻辑更为复杂的应用程序。

(3) 强调采用标签、表达式语言、JSTL(标准标签库)等替代脚本段代码。JSP 建议尽可能使用标签、表达式语言、标准标签库等 JSP 元素生成动态内容，以替代最初生成动态内容的脚本元素。

(4) 可移植性和安全性好。由于 JSP 最终都被转换成为基于 Java 的 Servlet，因此 JSP 页面就具有 Java 语言和 Servlet 技术的所有好处，包括可移植性、安全性、跨平台和跨 Web 服务器。

4.1.3 JSP 执行过程和第一次访问

1. JSP 执行过程

JSP 的执行过程如图 4-1 所示。

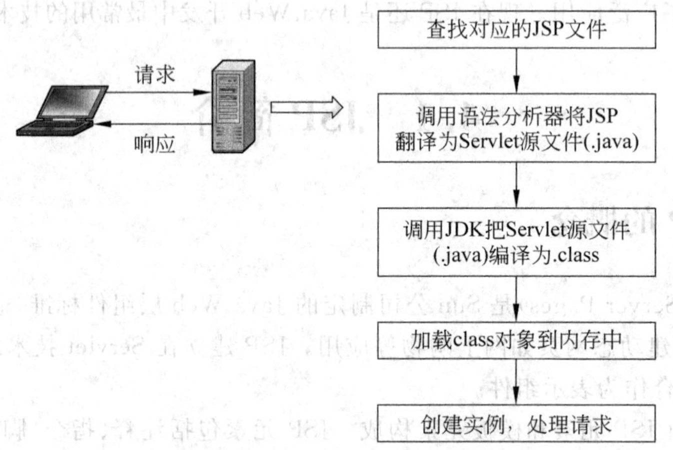

图 4-1 JSP 的执行过程

(1) 客户端(通常是浏览器)请求服务器上的某个 JSP 页面。

(2) 服务器接收请求后，根据请求查找服务器上对应的 JSP 页面。

(3) 服务器调用 JSP 语法分析器将该 JSP 翻译为 Servlet 程序(即.java 文件)。

(4) 服务器调用 JDK 将 Servlet 程序编译为字节码(.class 文件)并载入内存。

(5) 创建 Servlet 对象并以线程方式处理请求并向客户端返回响应。

2. 第一次访问 JSP 比较慢的原因

JSP 第一次被访问(或修改后第一次被访问)时，JSP 引擎会将 JSP 翻译为 Servlet 程序和编译为字节码文件、加载字节码文件到内存中、创建实例处理请求并返回响应。因此 JSP 在第一次被访问时，用户在客户端等待的时间较长。

但是如果第二次访问，JSP 引擎直接调用内存中的实例处理请求返回响应，响应的时间就快多了。在这种情况下与执行 Servlet 程序的速度几乎相同。

4.2 JSP 注释

在 JSP 中可以使用的注释有 4 种。

1. HTML 注释

语法格式：

```
<!--注释内容-->
```

例如，

```
<!--第一个注释是 HTML 注释，该类注释在客户端是可见 -->
```

2. HTML 结合 JSP 表达式的注释

语法格式：

```
<!--注释内容中包含 JSP 表达式-->
```

例如，

```
<!--第二个注释是 HTML 结合 JSP 表达式的注释<%=(new java.util.Date())
.toLocaleString() %>-->
```

该注释在注释内容中使用了 JSP 表达式来产生日期。

3. JSP 注释

语法格式：

```
<%--注释内容--%>
```

例如，

```
<%--第三个注释是 JSP 注释，该注释在客户端是不可见的--%>
```

4. Java 注释

在 JSP 页面可以使用脚本段即 Java 代码片段，这样可以使用 Java 的单行注释或多行注释。

```
<%
    //第四个注释是 Java 注释，即 Java 语言中的单行、多行注释
    /*
    User user =new User();
    */
```

```
%>
```

前两种注释会发送到客户端,即在浏览器中右击选择“源文件”命令,可以看到注释内容。第 3 种注释和第 4 种注释不会发送到客户端,在浏览器中右击选择“源文件”命令将不能看到注释内容。

4.3 JSP 指令元素

JSP 指令由 JSP 引擎执行,主要用来告诉 JSP 引擎如何处理 JSP 页面,并不会直接产生任何可见输出。

4.3.1 page 指令

page 指令用于指定 JSP 文件的全局属性,如页面编码格式、缓冲区大小、是否使用 session 等。

1. page 指令语法格式

page 指令的语法格式如下:

```
<%@page
[ language="java" ]
[extends="package.class" ]
[import="{package.class | package.* },..." ]
[session="true | false" ]
[contentType="mimeType[;charset=characterSet]"]
[pageEncoding="encodingType"]
[buffer="none | 8kb | sizekb" ]
[autoFlush="true | false" ]
[isThreadSafe="true | false" ]
[info="text" ]
[errorPage="relativeURL" ]
[ isErrorPage="true | false" ]
%>
```

page 指令的属性说明见表 4-1。

表 4-1 page 指令的属性说明

属　　性	说　　明	默认值	举　　例
language	指定 JSP 使用的脚本语言,目前只能是 java	java	language="java"

续表

属　性	说　明	默认值	举　例
extends	指定 JSP 页面产生的 Servlet 继承的父类。 一般不需要设置该属性	HttpJspBase	extends = "myPackage. myClass"
import	指定 JSP 需要导入的 Java 包或类的列表	java. lang. *, javax. servlet. *, javax. servlet. jsp. *, javax. servlet. http. *	import="java. io. *"
session	指定 JSP 是否使用 session 对象。 取值为 true 或 false	true	session="true"
contentType	指定 JSP 页面响应结果的 MIME 类型和字符编码类型	text/html;charset=ISO-8859-1	contentType="text/html;charset=GBK"
pageEncoding	指定 JSP 文件本身的字符编码类型	ISO-8859-1	pageEncoding="UTF-8"
buffer	设置 JSP 页面向客户端输出的内容是否使用缓冲区,或使用缓冲区时缓冲区大小。 取值为 none、8kb、自定义大小	8kb	buffer="none" buffer="8kb" buffer="64kb"
autoFlush	当使用缓冲区时,指定缓冲区填满时是否自动刷新,即缓冲区填满时自动向客户端输出缓冲区中的内容。 取值为 true,表示自动刷新;取值为 false,表示手动刷新。手动刷新时,若缓存填满后没有向客户端输出,则可能会出现严重的错误	true	autoFlush="true"
isThreadSafe	指定 JSP 页面是否支持多线程访问。取值为 true 或 false	true	isThreadSafe="true"
info	指定插入到 JSP 中的一段文本。在该 JSP 中通过 Servlet 的 getServletInfo()方法得到插入的文本	无	info="一段说明文字" String str = getServletInfo();
errorPage	指定当前 JSP 页面发生异常时,请求被转发到 errorPage 指定的页面	无	errorPage="error/error. jsp"
isErrorPage	指定当前页面是出错处理页面。在出错处理页面可以使用 exception 对象	false	isErrorPage="true"
isELIgnored	指定是否忽略 EL。 取值为 true,表示 Servlet 容器忽略对"${}"中表达式的计算	默认值由 Servlet 版本确定	isELIgnored="true"

2. page 指令使用注意事项

page 指令作用于整个 JSP 页面,包括静态包含的文件,即使用<%@ include %>指令包含的文件。但是 page 指令不能作用于动态的包含文件，如 <jsp:include>包含的文件。

page 指令实际使用时，只需对使用的属性进行设置而不是同时设置所有的属性。可以在一个页面中多次使用 page 指令，除了 import 属性外其他属性只能用一次。因为 import 属性类似于 Java 中的 import 语句。

page 指令可以放在 JSP 页面的任何位置。为了可读性，通常把它放在 JSP 文件的顶部。

page 指令的 contentType 属性与 pageEncoding 属性指定的字符编码类型要一致。

4.3.2 include 指令

include 指令用于包含另外一个资源。include 指令称为静态包含，指在翻译阶段将被包含文件与当前页面一起编译生成一个 Servlet 文件。

1. include 指令语法格式

include 指令语法格式如下：

```
<%@include file="fileUrl" %>
```

include 指令的属性说明见表 4-2。

表 4-2 include 指令的属性说明

属　性	说　明
file	指定被包含的文件的路径

被包含的文件的路径可以是相对路径或是绝对路径。如果被包含的文件是以文件名或目录名开头，那么路径就是相对路径；如果以"/"开头，那么路径就是绝对路径。

被包含的资源可以是 JSP 文件或者是 HTML 文件或者是文本文件或者是一段 Java 代码。

例如，如下的 include.jsp 使用 include 指令包含了 included.jsp。include.jsp 关键代码如下：

```
<!--include.jsp-->
<html>
...
<body bgcolor="white">
    <font color="blue">
        The current date and time are <%@include file="included.jsp" %>
        <br>执行完了 include 指令后执行本行。
    </font>
</body>
</html>
```

被包含文件 included.jsp 的作用是创建一个时间日期对象，其关键代码如下：

```
<!--included.jsp-->
<%@page import="java.util.*"%>
<%=(new java.util.Date()).toLocaleString()%>
```

在 include.jsp 中，程序执行到<%@ include file="included.jsp" %>时，会执行被包含的 included.jsp 文件，included.jsp 执行完毕，接着执行<%@ include file="included.jsp" %>后面的代码。

2. include 指令的实际应用

采用 JSP 进行实际项目开发的时候，在多个页面可能有相同的代码。为了提高可重用性，JSP 允许将需要多次使用的相同功能的代码封装在一个单独的 JSP 文件中。若其他页面需要使用这种功能，直接包含该页面即可。这就是 include 指令的作用。例如，某个网站的标题、页脚、导航栏是相同的，通常把这些做成单独的 JSP 页面，在其他页面使用 include 指令将它们包含进来。

4.3.3 taglib 指令

taglib 指令用于引入标签库，其语法格式如下：

```
<%@taglib uri="URIToTagLibrary" prefix="tagPrefix"%>
```

taglib 指令的属性说明见表 4-3。

表 4-3 taglib 指令的属性说明

属　　性	说　　明
uri	指定该标签库的位置
prefix	指定 JSP 页面中标签库的前缀

taglib 指令的详细使用说明见第 7 章。

4.4 脚本元素

脚本元素包括脚本段、表达式、声明。

1. 脚本段

脚本段就是 Java 代码片段。它不是完整的类而是类的一部分。语法格式如下：

```
<%Java code fragment%>
```

2. 表达式

表达式是符合 Java 语法的合法表达式。其作用在当前位置显示表达式的结果。语法格式如下：

```
<%=expression %>
```

3. 声明

声明是一段 Java 代码，用来声明变量或者方法或者类。声明后的变量、方法、类可在该 JSP 文件的任何地方使用。声明的变量属于实例变量。由于 JSP 是以多线程方式服务的，这意味着当多个用户在访问该网页时，将共享该变量。

声明的 JSP 语法格式如下：

```
<%! declaration; [ declaration; ] ... %>
```

例如，如下代码声明了一个变量和一个方法：

```
<body>
    <%--以下分别声明了一个实例变量和一个方法 --%>
    <%! int i=5; %>
    <%! int add(int a,int b){
            return a+b;
        }
    %>
    <%--使用表达式输出声明的实例变量--%>
    i=<%=i %><br>
    <%--使用脚本段调用声明的方法并用表达式输出结果 --%>
    <%
        int x=4,y=5,z;
        z=add(x,y);
    %>
    z=<%=z %>
</body>
```

4.5 动作元素

在 JSP 页面中有些符合 XML 语法的 JSP 标记，能够实现请求转发、包含动态文件、访问 JavaBean 等功能，这些标记称为 JSP 动作元素。

JSP 动作元素的一般语法格式有两种形式。第一种不带标记体，称为空标记，格式如下：

```
<prefix:tag attribute1=value1 attribute2=value2…/>
```

第二种带标记体称为非空标记，格式如下：

```
<prefix:tag attribute1=value1 attribute2=value2…>
  标记体
<prefix:tag />
```

JSP 常用的动作元素有<jsp:forward>、<jsp:include>、<jsp:useBean>、<jsp:getProperty>、<jsp:setProperty>等。在此主要介绍<jsp:forward>和<jsp:include>，后 3 种在 JSP 使用 JavaBean 时介绍。

4.5.1 <jsp:forward>

<jsp:forward>将请求转发到另一个资源，如 JSP 页面或者 Servlet 或者静态 HTML 页面。当 JSP 页面遇到<jsp:forward>时就停止执行当前的 JSP 转而执行另一个资源。

语法格式 1：

```
<jsp:forward page={"relativeURL" | "<%=expression %>"} />
```

语法格式 2：

```
<jsp:forward page={"relativeURL" | "<%=expression %>"} >
   <jsp:param name="parameterName"
      value="{parameterValue |<%=expression %>}" />
   [<jsp:param … />]
</jsp:forward>
```

<jsp:forward>的属性说明见表 4-4。

表 4-4 <jsp:forward>的属性说明

属　性	说　明
page	指定要转发到的资源的 URL，取值为一个表达式或是一个字符串

在语法格式 2 中，<jsp:forward>使用子标记<jsp:param>向转发到的 URL 传递一个或多个参数。

<jsp:param>的语法格式：

```
<jsp:param name="parameterName" value="{parameterValue | <%=expression %>}" />
```

<jsp:forward>的属性说明见表 4-5。

表 4-5 <jsp:forward>的属性说明

属　性	说　明
name	指定传递参数的名字
value	指定传递参数的值

<jsp:forward>可以有多个<jsp:param/>子标记，表示可以传递多个参数。传递参数时要以名值对形式进行；在转发到的 URL 可以通过 request 的 getParameter(String name)取得传递参数的值，这时该 URL 所代表的资源必须是动态文件，如 JSP 或 Servlet。

例如，如下的 original.jsp 使用<jsp:forward>从当前页面 original.jsp 将请求转发到另一个页面 destination.jsp。original.jsp 关键代码如下：

```
<body>
    <!--original.jsp-->
    <jsp:forward page="destination.jsp">
        <jsp:param name="userName" value="zhangsan"/>
    </jsp:forward>
</body>
```

destination.jsp 关键代码如下：

```
<!--destination.jsp-->
<%
    String useName=request.getParameter("userName");
    String outStr="欢迎你!";
    outStr+=useName;
    out.println(outStr);
%>
```

4.5.2 <jsp:include>

<jsp:include>称为动态包含，允许包含动态文件或静态文件。如果包含的文件是静态文件(如 HTML 页面)，则在翻译阶段将被包含文件与当前页面一起编译生成一个 Servlet 文件。如果包含的文件是动态文件(如 JSP 页面)，则在 JSP 最终作为 Servlet 运行时将被包含文件的响应结果包含进来。包含动态文件时当前 JSP 页面与被包含的 JSP 会生成两个 Servlet 文件。

语法格式 1：

```
<jsp:include page="{ relativeURL | <%=expression %>}"flush="true | false" />
```

语法格式 2：

```
<jsp:include page="{ relativeURL | <%=expression %>}"flush="true | false" >
    <jsp:param name="parameterName"
        value="{ parameterValue | <%=expression %>}" />
      [<jsp:param … />]
</jsp:include>
```

<jsp:include>的属性说明见表 4-6。

表 4-6 <jsp:include>的属性说明

属　性	说　明
page	指定被包含文件的 URL,取值为字符串或表达式
flush	指定在读入包含内容之前是否清空任何现有的缓冲区,取值为 true 或 false,JSP 1.2 默认值为 false

在语法格式 2 中,<jsp:include>使用一个或多个子标记<jsp:param>传递一个或多个参数到指定的动态文件。

4.6 内建对象

为了简化 JSP 页面开发的复杂性和提高开发效率,JSP 提供了一些对象用于向客户端输出数据、处理客户端的请求信息、设置响应等。这些对象不需要 JSP 开发者实例化,它们是由容器创建和管理的,称为内建对象。所有的内建对象只能用于脚本段代码或表达式,不能在 JSP 声明中使用。

4.6.1 out 对象

out 对象是 javax.servlet.jsp.JspWriter 类的实例。在 JSP 中,out 对象用于向客户端输出数据和管理响应缓冲。

out 对象的常用方法说明见表 4-7。

表 4-7 out 对象的常用方法说明

方　法	方法说明
out.print(Type)	输出数据。Type 可以是基本数据类型(boolean、char、float、double、byte、short、int、long)、String 类型、Object 类型
out.println(Type)	输出数据后会再后面加上一个换行符。Type 同上
newline()	输出一个换行符号
flush()	输出缓冲区中的数据
close()	关闭输出流,清除缓冲区中的数据
clearBuffer()	清除缓冲区中的数据,并把数据写到客户端
clear()	清除缓冲区中的数据,不把缓冲区中的数据写到客户端
getBufferSize()	得到缓冲区大小(单位是字节)。缓冲区大小可用<%@ page buffer="Size"%>设置
isAutoFlush()	返回布尔值,如果是自动刷新返回 true,否则返回 false
getRemaining()	得到缓冲区没有使用的空间的大小(单位是字节)

out 对象的两种重要方法是 out. print(Type)或 out. println(Type)。第一种方法只是输出数据；第二种方法输出数据后会再后面加上一个换行符，但是该换行符不能被浏览器解析，真正换行时需要使用 out. println("<br>")来实现。

例如，如下代码使用 out 对象输出了 string 对象、boolean 对象和 integer 对象。

```
<body>
    <%
        out.println("输出 String");
        out.println(true);
        out.println(123);
    %>
</body>
```

4.6.2 response 对象

response 对象是 javax. servlet. http. HttpServletResponse 接口的实例。response 对象代表服务器向客户端发送的响应。

response 对象的常用方法说明见表 4-8。

表 4-8 response 对象的常用方法说明

方　法	方法说明
setHeader(String name,String value)	设置 HTTP 的头信息，参数 name 指定头信息名，参数 value 指定头信息的值
sendRedirect(Stringurl) throws IOException	重定向到 url 指定的路径
addCookie(Cookie cookie)	向客户端添加 Cookie
setContentType(String type)	设置内容的返回类型

response 对象的主要用途是设置 HTTP 的头信息、重定向、添加 Cookie。

1. 设置 HTTP 头信息

语法格式：

```
response.setHeader(String name,String value)
```

设置 HTTP 头信息中最有用的一个是 refresh，可以用于当前页面刷新或访问另一个页面。

例如，以下 refresh. jsp 设置头信息进行页面刷新，该页面每隔一秒将自动刷新一次。

```
<body>
  <%! int i=0; %>
  <%
    //每秒钟刷新一次，每次使得 i 加 1
```

```
    response.setHeader("refresh","1");
  %>
  <h1><%=i++%></h1>
</body>
```

如下代码表示过了 2s 会访问另一个页面 useOut. jsp。

```
<body>
<%//2s 后跳转到 useOut.jsp
    response.setHeader("refresh","2;URL=useOut.jsp");
%>
</body>
```

2. 重定向

语法格式：

```
response. sendRedirect(String url)
```

重定向属于客户端跳转，即客户端重新向服务器发送请求，原来的请求丢失。<jsp:forward>属于服务器端的跳转，原来的请求不会丢失。

例如有两个页面：redirect1. jsp，redirect2. jsp。在 redirect1. jsp 中使用 request 中保存一个 String 的对象"Mike"后再重定向到 redirect2. jsp，在 redirect2. jsp 中尝试取得 name 将得到空。

redirect1. jsp 代码如下：

```
<body>
    <%
        request.setAttribute("name","Mike");
        response.sendRedirect("redirect2.jsp");
    %>
</body>
```

redirect2. jsp 代码如下：

```
<body>
    查看能否接收到 redirect2.jsp 中的请求参数 name<br>
    name=<%=request.getAttribute("name")%>
</body>
```

3. 添加 Cookie

语法格式：

```
response.addCookie(Cookie cookie)
```

例如，如下代码向客户端增加两个 Cookie。

```
<%
  // 准备两个 Cookie
  Cookie c1 =new Cookie("username","Mike") ;
  Cookie c2 =new Cookie("userpass","123") ;
  // 通过 response 添加两个 Cookie 到客户端
  response.addCookie(c1) ;
  response.addCookie(c2) ;
%>
```

获取客户端全部 Cookie 使用 request. getCookie()方法。例如,如下代码遍历全部 Cookie 获得上述添加的两个 Cookie。如果浏览器禁用了 Cookie,将导致对 Cookie 的操作失败。

```
<%
  // 取得设置的 Cookie 对象
  Cookie c[] =request.getCookies() ;
  System.out.println(c) ;
  for(Cookie c : cookies){
      if(c.getName().equals("username")){
          break;
      } else if (c.getName().equals("userpass")){
      break;
    }
%>
  <%=c.getName()%>--><%=c.getValue()%>
<%
  }
%>
```

4.6.3 request 对象

request 对象是 javax. servlet. http. HttpServletRequest 接口的实例。客户端的请求信息被 Servlet 容器封装到该对象中。

request 对象的常用方法说明见表 4-9。

表 4-9 request 对象的常用方法说明

方　　法	方法说明
setAttribute(String name,Object obj)	在请求作用域 request 中保存一个变量,变量名为 name,变量值为 obj
getAttribute(String name)	返回请求作用域 request 作用域中 name 的值
removeAttribute(String name):	删除请求作用域 request 中的 name 变量

续表

方　法	方法说明
setCharacterEncoding(String encoding)	设置请求的字符编码类型
getParameter(String name)；	获得请求参数 name 的值
getParameterNames()	获得全部参数的名字，结果是一个枚举的实例
getParameterValues(String name)；	获得请求参数名为 name 的全部值

例如，如下 useRequest.jsp 中列出了 request 对象的部分常用方法。

```
<body>
<%
    // 设置请求编码为 GBK
    request.setCharacterEncoding("GBK");
    // 获得请求参数 name 的值
    out.println(request.getParameter("name"));
    out.print("<br>");
    // 获得服务器的端口号
    out.println(request.getServerPort());
    out.print("<br>");
    //将一个对象放在 request 作用域
    String name="guoqing";
    request.setAttribute("userName",name);
    //从 request 中取出并打印输出该对象
    String userName=(String)request.getAttribute("userName");
    out.print("<br>");
    out.println("取出 request 对象中的变量 userName 是:"+userName);
%>
</body>
```

4.6.4　session 对象

1. 用户会话跟踪

HTTP 协议是无状态的，这表示当用户请求一个资源，服务器接收和处理请求、返回响应到客户端后，就关闭了连接。这样前一个请求结束而另一个请求开始时，HTTP 协议无法记录前一个请求的信息。

Web 应用常常需要处理用户的多个连续的请求并且需要记住前一次请求的信息，这就是用户会话跟踪。例如网上书店系统中的购物车，当某个客户请求将商品放入购物车时，Web 服务器必须根据客户上次请求信息中的客户身份，将商品放入客户对应的购物车中。

Web 应用中实现用户会话跟踪的方式有 Cookie、隐藏表单域、URL 重写、session。

1）Cookie

Web 服务器可以将唯一的会话 ID 保存在 Cookie 中，并将 Cookie 发送到每个 Web 客户端；Web 服务器对后续的请求处理时取得 Cookie 进行识别。

由于浏览器可能不支持 Cookie 或可以禁用 Cookie，因此使用 Cookie 不支持常规的会话跟踪。

2）隐藏表单域

Web 服务器可以将唯一的会话 ID 保存在 HTML 隐藏表单域中，并将该隐藏表单域发送到客户端。

例如，<input type = "hidden" name = "sessionid" value = "123456">，客户端提交表单后，Web 服务器获得会话 ID 可以实现用户会话跟踪。

这种方式仅限于每个页面都是表单提交且表单是动态生成的情况。单击常规超链接(<a href="...">)不会产生表单提交，因此隐藏表单域也不支持常规会话跟踪。

3）URL 重写

Web 服务器在发送到客户端每个 URL 的末尾附加会话 ID，在 Web 服务器上获得该会话 ID。

例如，http://www.sun.com/file.html;sessionid=123456，Web 服务器可以获得会话 ID 可以实现用户会话跟踪。

URL 重写解决了浏览器不支持 Cookie 时维护会话的问题。但是这种方式的缺点是必须动态生成每个 URL 来分配会话 ID，无法用于页面是静态 HTML 页面的情况。

4）session

session 是各种 Web 开发技术提供的用户会话跟踪解决方案。当某个客户端访问 Web 应用时，Web 服务器会为该客户端创建一个 session 对象。该 session 对象具有唯一的 String 类型的 ID(会话 ID)。如果客户端允许使用 Cookie，那么 Web 服务器会将该 ID 放入客户端的 Cookie 中；如果客户端不允许使用 Cookie，那么服务器会将该 ID 以 URL 重写的方式发送到客户端。这样就建立了 session 对象和客户之间的一一对应的关系。

2. session 对象

session 对象是 javax.servlet.http.HttpSession 接口的实例。session 对象的常用方法说明见表 4-10。

表 4-10 session 对象的常用方法说明

方 法	方法说明
void setAttribute(String name, Object value)	在 session 中保存一个对象 value，参数 name 指定对象 value 的名字；在 session 中如果 name 已经存在，那么 name 对应的对象被替换成 value 对象。
Object getAttribute(String name)	从 session 中取出参数 name 代表的对象；如果 name 不存在，则返回 null
void removeAttribute(String name)	删除 session 中 name 指定的对象。如果 name 不存在，那么该方法不进行任何操作

续表

方　　法	方法说明
void invalidate()	使 session 失效，同时删除 session 保存的所有属性对象
String getId()	返回 session 的 ID
void setMaxInactiveInterval(int interval)	设置 session 的有效时间，单位为秒(s)。该时间是客户请求之间的最长时间间隔；如果请求之间的时间超过该有效时间，则 JSP 容器认为请求属于两个不同的会话
void getMaxInactiveInterval()	返回 session 的有效时间，单位为秒(s)
Boolean isNew()	判断当前 session 是否是新创建的 session

下面介绍在用户登录中使用 session。login.jsp 页面用于客户输入用户名和密码，其关键代码如下：

```
<body>
    <h3>用户登录</h3>
    <form action="login_do.jsp" method="post">
        用户名:<input type="text" name="name" width="20"/><br/>
        密码:<input type="password" name="password" width="20"/><br/>
        <input type="submit" value="登录"/>
    </form>
</body>
```

输入用户名和密码后单击"登录"按钮，用户提交的信息交由 login_do.jsp 处理，其关键代码如下：

```
<body>
<%
    String name=request.getParameter("name");
    String password=request.getParameter("password");
    if(name.equals("javaweb")&&password.equals("123")){
        session.setAttribute("name",name);
        request.getRequestDispatcher("index.jsp").forward(request,response);
    }else{
        response.sendRedirect("login.html");
    }
%>
</body>
```

login_do.jsp 首先用于接收请求参数用户名和密码，接着判断如果登录用户名和密码分别是 javaweb 和 123，在 session 中保存登录用户的用户名后跳转到 index.jsp 页面；否则跳转到登录页面 login.jsp。

index.jsp 页面用户显示登录成功并输出登录的用户名，其关键代码如下：

```
<body>
```

```
欢迎您!<%=session.getAttribute("name") %>,登录成功!
</body>
```

4.6.5 application 对象

application 对象是 javax.servlet.ServletContext 接口的实例。application 对象用于保存同一个 Web 应用的所有页面和所有用户的共享数据。每当 Web 服务器启动时，Servlet 容器为每个应用创建一个 application 对象；当服务器关闭的时候，application 对象就会被销毁。

application 对象的常用方法说明见表 4-11。

表 4-11　application 对象的常用方法说明

方　法	方法说明
String getRealPath(String path)	返回参数 path 的绝对路径
void setAttribute(String name,Object obj)	将对象 obj 保存在 application 中，name 代表 obj 的名字
Object getAttribute(String name)	返回 application 中 name 指定的对象
void removeAttribute(String name)	删除 application 中 name 指定的对象

例如，如下代码演示了 application 对象的常用方法。

```
<body>
<%
    out.println("contextPath="+application.getContextPath()+"<br/>");
    application.setAttribute("number",new Integer(6));
    out.println("number="+application.getAttribute("number")+"<br/>");
    application.removeAttribute("number");
    out.println("number="+application.getAttribute("number")+"<br/>");
    out.println("realpath="+application.getRealPath("/ch06/index.jsp")+"<br/>");
%>
</body>
```

4.6.6 pageContext 对象

pageContext 对象是 javax.servlet.jsp.PageContext 类的实例。pageContext 对象主要用于管理和访问其他内建对象。

pageContext 对象的常用方法说明见表 4-12。

表 4-12　pageContext 对象的常用方法说明

方　法	方法说明
JspWriter getOut()	返回当前页面的输出流对象 out

续表

方　法	方法说明
ServletRequest getRequest()	返回当前页面的请求对象 request
ServletResponse getResponse()	返回当前页面的响应对象 response
ServletContext getServletContext()	返回当前页面的上下文对象 context
HttpSession getSession()	返回当前页面的会话对象 session
Object getAttribute(String name)	返回 page 作用域中 name 指定的对象
Object getAttribute(String name, int scope)	返回 scope 指定的作用域中 name 指定的对象
int getAttributesScope(String name)	返回 name 代表的对象所属的作用域
void removeAttribute(String name)	删除 page 作用域中 name 指定的对象
void removeAttribute(String name, int scope)	删除 scope 指定的范围中 name 指定的对象
void setAttribute(String name, Object obj)	将对象 obj 保存到 page 作用域，name 代表 obj 的名字
void setAttribute(String name, Object obj, int scope)	将对象 obj 保存到 scope 指定的作用域中，name 代表 obj 的名字

pageContext 对象的保存、删除、获得属性对象的方法与前面内容中 request、session、application 对象的对应方法相比多了一个表示作用域的参数。

例如，如下代码使用 request 对象把一个对象保存在 request 范围中：

```
request.setAttribute("name", "Mike");
```

使用 pageContext 对象完成等价功能的代码如下：

```
pageContext.setAttribute("name", "Mike",pageContext.REQUEST_SCOPE);
```

将对象保存在 page 作用域中使用 pageContext. setAttribute(String name, Object obj, pageContext. PAGE_SCOPE)方法。

4.6.7 config 对象

config 对象是 javax. servlet. ServletConfig 接口的实例，它代表 Servlet 的配置信息。JSP 页面最终要化为 Servlet 运行，因此在 web. xml 配置 JSP 为 Servlet 后在页面可以访问配置信息。

config 对象的常用方法说明见表 4-13。

表 4-13　config 对象的常用方法说明

方法名	方法说明
ServletContext getServletContext()	返回 Servlet 上下文对象 ServletContext
String getInitParameter(String name)	返回 name 指定的初始化参数

续表

方法名	方法说明
Enumeration getInitParameterNames()	返回 Servlet 或 JSP 所有的初始化参数，返回类型是 Enumeration
String getServletName()	返回 Servlet 的名字

例如，useConfig. jsp 在 web. xml 中配置后，使用 config 对象获取配置信息。useConfig. jsp 的关键代码如下：

```
<body>
访问数据库的用户名为：
<%=config.getInitParameter("dbUserName") %><br/>
访问数据库的密码为：
<%=config.getInitParameter("dbUserPassword") %><br/>
数据库的 URL 为：
<%=config.getInitParameter("dbURL") %>
</body>
```

在 web. xml 中的配置 useConfig. jsp 的关键代码如下：

```
<servlet>
    <servlet-name>useConfig</servlet-name>
    <jsp-file>/ch04/useConfig.jsp</jsp-file>
    <init-param>
        <param-name>dbUserName</param-name>
        <param-value>sa</param-value>
    </init-param>
    <init-param>
        <param-name>dbUserPassword</param-name>
        <param-value>sa</param-value>
    </init-param>
    <init-param>
        <param-name>dbURL</param-name>
        <param-value>jdbc:sqlserver://localhost:1433;DataBaseName=myDB
        </param-value>
    </init-param>
</servlet>
<servlet-mapping>
    <servlet-name>useConfig</servlet-name>
    <url-pattern>/ch04/useConfig.html</url-pattern>
</servlet-mapping>
```

4.6.8 exception 对象

exception 对象是 java.lang.Throwable 类的实例，它表示的是 JSP 页面运行时出现的运行时错误。

exception 对象的常用方法说明见表 4-14。

表 4-14 exception 对象的常用方法说明

方　　法	方法说明
String getMessage()	返回异常信息
String toString()	返回异常对象的字符串描述
void printStackTrace()	打印异常及其堆栈轨迹

例如，在如下的 divide.jsp 中，使用整数除 0 错误演示如何使用 exception 对象。在可能会出现错误的 divide.jsp 中，要设置 page 指令的 errorPage 属性为 error.jsp，这样当页面 divide.jsp 出现错误时，会跳转到 error.jsp。divide.jsp 的关键代码如下：

```
<%@page language="java" contentType="text/html; charset=UTF-8"
        pageEncoding="UTF-8" errorPage="error.jsp"%>
<html><head><title>divide</title></head>
<body>
<%
    int z=5/0;
%>
</body>
</html>
```

error.jsp 的关键代码如下：

```
<%@page language="java" contentType="text/HTML; charset=UTF-8"
    pageEncoding="UTF-8" isErrorPage="true"%>
<html>
<head><title>error</title></head>
<body>
    divide.jsp 发生了以下错误:<br/>
    <%=  exception.getMessage()%>
</body>
</ html>
```

小　　结

本章主要介绍了 JSP 的构成，JSP 的注释、指令元素、脚本元素、动作元素、内建对象等基础语法知识。JSP 基础语法比较琐碎，掌握起来有一定困难。

在基于 MVC 的应用中，JSP 充当表示层组件，因此 JSP 使用较多的语法包括指令元素、脚本元素中的表达式、动态包含<jsp:include>、少数内建对象等；很多语法，如脚本段、大部分内建对象等可以使用表达式语言、JSTL 替代。请求转发的<jsp:forward>通常在 Servlet 中进行。

思考与习题

1. 什么是 JSP?
2. 画图并以文字描述 JSP 执行过程。
3. 第一次访问 JSP 为什么比较慢?
4. JSP 注释共有几种? 举例说明。
5. 写出 page 指令的 4 个常见属性及其含义。
6. 举例说明 include 指令的用法和含义。
7. 举例说明脚本段的、表达式、声明的用法。
8. 举例说明动作元素 forward 和 include 的用法及其含义。
9. 举例 JSP 的相关内建对象的使用及其含义。

第5章 JavaBean

JavaBean 是一种可重用的、跨平台的软件组件模型。在 Java 应用中的 JavaBean 有两种：第一种是用在用户界面如 AWT 或 Swing 中的 JavaBean；第二种是用在非用户界面的 JavaBean。在 JSP 中使用的是第二种。

5.1 JavaBean 规范

(1) JavaBean 是一个 public 的类。即该类具有 public 的访问权限。

(2) JavaBean 必须有一个无参数的构造方法。一个 JavaBean 可以有一个或有多个构造方法，但是必须有一个无参数的构造方法。

(3) JavaBean 的属性一般是非 public 的，访问属性应该通过 public 的 setter 和 getter 方法。

(4) JavaBean 中可以封装业务逻辑方法，包括进行数据库访问操作等。

例如，如下的 User 类满足 JavaBean 规范，因此可以称 User 类是一个 JavaBean。

```
public class User {
    private int id;
    private String name;
    private String password;
    private int age;
    public int getId() {
        return id;
    }
    public void setId(int id) {
        this.id =id;
    }
    public String getName() {
        return name;
    }
    public void setName(String name) {
        this.name =name;
```

```
    }
    public String getPassword() {
        return password;
    }
    public void setPassword(String password) {
        this.password =password;
    }
    public int getAge() {
        return age;
    }
    public void setAge(int age) {
        this.age =age;
    }
    public void addAge(){
        this.age++;
    }
}
```

5.2 访问 JavaBean

5.2.1 使用脚本段代码访问 JavaBean

1. 使用 page 指令导入 JavaBean

使用 page 指令的 import 属性导入 JavaBean。例如

```
<%@page import="ch05.User" %>
```

2. 使用脚本段代码访问 JavaBean

```
<body>
<%
    User user =new User();
    user.setName("Mike") ;
    user.setPassword("123");
%>
用户名:<%=user.getName() %><br/>
密码:<%=user.getPassword() %>
</body>
```

以上代码首先创建 JavaBean 对象 user,接着调用 setName()方法和 setPassword()方法为其属性赋值,分别是 Mike 和 123;最后使用 JSP 表达式在页面显示用户名和密码。

5.2.2 使用动作元素访问 JavaBean

1. <jsp:useBean>

<jsp:useBean>语法格式如下：

```
<jsp:useBean id="beanName" class="package.className"
    scope="page|request|session|application" >
</jsp:useBean>
```

表示创建 package.className 类的实例 beanName，并将该实例置于 scope 属性指定的某个作用域。实际上，Servlet 容器首先在 scope 指定的作用域中查找 beanName 实例。如果找到，就可以在 JSP 页面使用该实例；如果找不到，Servlet 容器才创建一个新的实例。

<jsp:useBean>的常用属性说明见表 5-1。

表 5-1 <jsp:useBean>的常用属性说明

属　性	说　明
id	指定 JavaBean 实例名
class	指定 JavaBean 实例所属类的全限定名，即包名.类名
scope	指定 JavaBean 属性范围(page|request|session|application)

2. <jsp:getProperty>

<jsp:getProperty>的语法格式如下：

```
<jsp:getproperty name="beanName" property="propertyName" />
```

表示读取并显示一个 JavaBean 实例 beanName 的 propertyName 属性值。

<jsp:getProperty>属性说明见表 5-2。

表 5-2 <jsp:getProperty>属性说明

属　性	说　明
name	指定 JavaBean 实例名
property	指定待读取的 JavaBean 属性

3. <jsp:setProperty>

<jsp:setProperty>的语法格式有如下 4 种：

```
<jsp:setpropertyname="beanName" property="*" />|
<jsp:setpropertyname="beanName" property="propertyName" />|
<jsp:setpropertyname="beanName" property=" propertyName" param="paramName"/>|
<jsp:setpropertyname="beanName" property=" propertyName" value="expression"/>
```

表示设置一个 JavaBean 实例的属性值，name 指定的 JavaBean 实例的名字，property 属性指定 JavaBean 实例的属性。

<jsp:setProperty>属性说明见表 5-3。

表 5-3　<jsp：setProperty>属性说明

属　　性	说　　明
name	指定 JavaBean 实例名
property	指定 JavaBean 属性
param	指定使用请求参数设置 JavaBean 的属性
value	指定使用一个表达式设置 JavaBean 属性。该表达式可以是某个请求参数，或是某个固定值等

(1) <jsp:setproperty name="beanName" property=" * " />

设置与请求参数中同名的 JavaBean 对象属性值并自动将请求参数转成 JavaBean 对应属性的类型；如果某个请求参数名与 JavaBean 的属性不同，那么不会设置该 JavaBean 的属性。

例如，用户注册时将请求参数用户名和年龄分别设置到 user 对象的属性 name 和 age 中。用户注册页面 register. html 的关键代码如下：

```
<body>
  用户注册<br/>
  <form action="register_do.jsp" method="post">
      用户名:<input type="text" name="name"/><br/>
      年龄:<input type="text" name="age"/><br>
      <input type="submit" value="注册"/>
  </form>
</body>
```

另一个页面 register_do. jsp 使用<jsp:useBean>实例化一个 JavaBean user，使用<jsp:setProperty>将请求参数用户名和年龄设置到 user 的对应属性中，并显示用户名和密码，其关键代码如下：

```
<body>
<jsp:useBean id="user" class="ch05.User" scope="page" />
<jsp:setProperty name="user" property=" * " />
显示用户注册信息<br/>
您的用户名:<jsp:getProperty property="name"name="user"/><br/>
您的年龄:<jsp:getProperty property="age" name="user"/>
</body>
```

第一个页面中的两个文本框的 name 和 age 与第二个页面中使用的 JavaBean 对象 user 的属性完全一致。

```
public class User {
```

```
    private int id;
    private String name;
    private String password;
    private int age;
    …
    }
```

第二个页面使用<jsp:useBean>实例化一个 JavaBean user，并使用<jsp:setProperty name="user" property="*" />将请求参数用户名 name 和年龄 age 来设置 JavaBean user 同名的 name 和 age 属性，最后使用<jsp:getProperty>分别读取和显示 JavaBean user 的 name 和 age 属性。

User 类的属性 age 是 int 类型，而页面传来的请求参数均为字符串。<jsp:setProperty name="user" property="*" />会将接收到的请求参数转换成 int 类型再设置为 user 的 age 属性。

(2) <jsp:setproperty name="beanName" property="propertyName" />

将请求参数自动转换类型并设置到 JavaBean 的同名属性；每次设置一个 JavaBean 属性，如果需要设置多个属性，就需要使用多次。

用户注册页面 register.html 的关键代码如下：

```
<body>
  用户注册<br/>
  <form action="register_do.jsp" method="post">
     用户名:<input type="text" name="name"/><br/>
     电子邮箱:<input type="text" name="age"/><br/>
     <input type="submit" value="注册"/>
  </form>
</body>
```

页面 register_do.jsp 的关键代码如下：

```
<body>
<jsp:useBean id="user" class="ch05.User" scope="page" />
<jsp:setProperty name="user" property="name" />
<jsp:setProperty name="user" property="age" />
显示用户注册信息<br/>
您的用户名:<jsp:getProperty property="name"name="user"/><br/>
您的年龄:<jsp:getProperty property="age"name="user"/>
</body>
```

register_do.jsp 首先实例化 JavaBean user，接着使用两次<jsp:setProperty>分别将接收到的请求参数 name 和 age 设置为 user 的同名属性，最后使用<jsp:getProperty>分别读取和显示 JavaBean user 的 name 和 age 属性。

(3) <jsp:setProperty name="beanName" property=" propertyName"
        param= "paramName"/>

用于当请求参数名与 JavaBean 的属性名不一致时，把请求参数自动进行类型转换并设置到 JavaBean 的属性；每次设置一个 JavaBean 属性，如果需要设置多个属性，就需要使用多次。

注册页面 register. html 的关键代码如下：

```
<body>
   用户注册<br/>
   <form action="register_do.jsp" method="post">
      用户名:<input type="text" name="username"/><br/>
      年龄:<input type="text" name="userage"/><br>
      <input type="submit" value="注册"/>
   </form>
</body>
```

第二个页面 register_do. jsp 的关键代码如下：

```
<body>
   <jsp:useBean id="user" class="ch05.User" scope="page" />
   <jsp:setProperty name="user" property="name" param="username"/>
   <jsp:setProperty name="user" property="age" param="userage"/>
   显示用户注册信息<br/>
   您的用户名:<jsp:getProperty property="name"name="user"/><br/>
   您的年龄:<jsp:getProperty property="age"name="user"/>
</body>
```

第二个页面 register_do. jsp 首先实例化 JavaBean user，接着使用如下语句：

```
<jsp:setProperty name="user" property="name" param="username"/>
<jsp:setProperty name="user" property="age" param="userage"/>
```

分别将接收到的请求参数 username 和 userage 设置为 user 对象的 name 和 age 属性。

(4) <jsp:setProperty name="beanName" property=" propertyName"
value= "expression"/>

使用 value 指定的表达式设置 JavaBean 的属性 propertyName；每次设置一个 JavaBean 属性，如果需要设置多个属性，就需要使用多次。

第一个页面 register. html 的关键代码如下：

```
<body>
   用户注册<br/>
   <form action="register_do.jsp" method="post">
      用户名:<input type="text" name="name"/><br/>
      电子邮箱:<input type="text" name="age"/><br/>
      <input type="submit" value="注册"/>
   </form>
</body>
```

第二个页面 register_do.jsp 的关键代码如下：

```
<body>
  <jsp:useBean id="user" class="ch05.User" scope="page" />
  <jsp:setProperty name="user" property="name" value="Mike" />
  <jsp:setProperty name="user" property="age"  value=${param.age}/>
  显示用户注册信息<br/>
  您的用户名:<jsp:getProperty property="name"name="user"/><br/>
  您的年龄:<jsp:getProperty property="age"name="user"/>
</body>
```

第二个页面 register_do.jsp 首先实例化 JavaBean user，接着使用<jsp:setProperty>将 user 的 name 属性设置为固定值 Mike，使用表达式语言将接收到的请求参数 age 设置为 user 的 age 属性，最后使用<jsp:getProperty>分别读取和显示 user 的 name 和 age 属性。

5.3 JSP+JavaBean 开发模式

在早期 JSP 编程中，JSP 页面包含了业务逻辑、流程控制逻辑和表示逻辑。这使得 JSP 页面维护和扩展困难。后来出现的 JSP+JavaBean 开发模式，JSP 负责显示逻辑和流程控制逻辑，JavaBean 负责封装业务逻辑。

本节使用 JSP+JavaBean 开发模式来实现用户注册功能。进行应用程序开发，需要进行需求分析、系统设计、详细设计和实现(编码)、测试等过程。注册案例比较简单，此处简化了相关过程。

1. 创建数据库和 user 表

在 MySQL 中创建数据库 testDB 和表 user。表 user 的字段包括 id、name、password、age，分别是 int、varchar、varchar、int 类型。

2. 创建 Eclipse 的 Web 项目、添加 JDBC 驱动程序、创建数据库帮助类

创建 Eclipse Java Web 项目 ch05-1；配置该项目构建路径添加 MySQL JDBC 驱动程序；数据库帮助类使用第 2 章的 DBConnection。

3. 创建 JavaBean User 并增加业务逻辑方法 register()

User 类的关键代码如下：

```
package ch05;
...
public class User {
    private int id;
```

```java
    private String name;
    private String password;
    private int age;
    …
    //返回注册成功的记录的 id
    public int register() {
        Connection conn =null;
        PreparedStatement ps =null;
        ResultSet rs =null;
        try {
            conn =DBConnection.getConnection();
            String sql ="insert into user(name,password,age)values(?,?,?)";
            ps =conn.preparedStatement(sql,
                Statement.RETURN_GENERATED_KEYS);
            ps.setString(1, this.name);
            ps.setString(2, this.password);
            ps.setInt(3, this.age);
            ps.executeUpdate();
            rs =ps.getGeneratedKeys();
            int id =-1;
            if (rs !=null && rs.next()) {
                id =rs.getInt(1);
            }
        } catch (SQLException e) {
            e.printStackTrace();
        }
        return id;
    }
    //根据 id 获得 User 对象
    public User getUserById(int id) {
        Connection conn =null;
        PreparedStatement ps =null;
        ResultSet rs =null;
        User user =null;
        try {
            conn =DBConnection.getConnection();
            String sql ="select * from user where id =?";
            ps =conn.preparedStatement(sql);
            ps.setInt(1, id);
            rs =ps.executeQuery();
            if (rs.next()) {
                user =new User();
                user.setId(rs.getInt("id"));
                user.setName(rs.getString("name"));
```

```
                user.setPassword(rs.getString("password"));
                user.setAge(rs.getInt("age"));
            }
        } catch (SQLException e) {
            e.printStackTrace();
        }
        return user;
    }
}
```

4. 创建有关 JSP 页面

分别创建页面 register. html、register_do. jsp、registerSuccess. jsp、registerFail. jsp。

- register. html：用于提供用户填写注册信息。
- register_do. jsp：处理 register. html 提交的注册信息，实现流程控制和业务逻辑调用。
- registerSuccess. jsp：在注册成功后显示注册成功的用户信息。
- registerFail. jsp：在注册失败后显示注册失败的提示。

register. html 页面的关键代码如下：

```
用户注册<br/>
<form action="register_do.jsp" method="post">
    用户名:<input type="text" name="name"/><br/>
    密码:<input type="password" name="password"/><br>
    年龄:<input type="text" name="age"/><br>
    <input type="submit" value="注册"/>
</form>
```

register_do. jsp 页面的关键代码如下：

```
<body>
<jsp:useBean id="user" class="ch05.User" scope="session" />
<jsp:setProperty name="user" property="*" />
<%
    int id =user.register();
    if(id >0){
        // user=user.get(1);
        request.getRequestDispatcher("registerSuccess.jsp").forward(
                request,response);
    }else{
                request.getRequestDispatcher("registerFail.jsp").forward(
                request,response);
    }
%>
</body>
```

registerSuccess.jsp 页面的关键代码如下：

```
<body>
<jsp:useBean id="user" class="ch06.User" scope="session"></jsp:useBean>
显示用户注册信息<br/>
用户名:<jsp:getProperty property="name" name="user"/><br/>
密码:<jsp:getProperty property="password" name="user"/><br/>
年龄:<jsp:getProperty property="age" name="user"/><br/>
</body>
```

在 register_do.jsp 中，以下代码：

```
<jsp:useBean id="user" class="ch06.User" scope="session" />
```

创建的 user 是 session 范围，所以在 registerSuccess.jsp 中，以下代码：

```
<jsp:useBean id="user" class="ch06.User" scope="session"></jsp:useBean>
```

并不会创建新的 JavaBean User 的实例 user，而是会继续使用 session 范围已经存在的 user。

registerFail.jsp 页面的关键代码如下：

```
<body>
注册失败！请联系管理员。
</body>
```

5.4 JSP+Servlet+JavaBean 开发模式

JSP+Servlet+JavaBean 是对 JSP+JavaBean 开发模式的改进。该模式中，JSP 负责显示逻辑，Servlet 负责流程控制逻辑，JavaBean 负责封装业务逻辑。该模式适合大型应用程序开发，便于应用程序的可维护性和可扩展性。

下面使用 JSP+Servlet+JavaBean 开发模式来实现用户列表功能。用户通过地址栏输入 http://localhost:8080/ch05-2/listAll 后，ListAllServlet 负责调用 JavaBean user 的 listAll()方法，获取数据库 testDB 中表 user 的全部记录并跳转 listAll.jsp 页面显示全部用户记录。

ListAllServlet 的关键代码如下：

```
package servlet;
…
@WebServlet("/listAll")
public class UserServlet extends HttpServlet {
    protected void doGet(HttpServletRequest request, HttpServletResponse response)
            throws ServletException, IOException {
        User user =new User();
```

```
        List<User>all =user.listAll();
        request.setAttribute("all", all);
        request.getRequestDispatcher("/listAll.jsp").forward(request, response);
    }
}
```

listAll.jsp 页面的关键代码如下：

```
<%@page language="java" contentType="text/html; charset=UTF-8"
    pageEncoding="UTF-8" import="java.util.*,ch06.User"%>
...
<body>
用户列表<br>
<table border="1">
<tr><td>用户名</td><td>年龄</td></tr>
<%
  List<User>all = (List<User>)request.getAttribute("all");
  if (all.size() >0) {
      for (User u : all) {
%>
      <tr><td><%=u.getName() %></td><td><%=u.getAge() %></td></tr>
<%} %>
<%
  }else {
%>
  <tr><td>没有任何记录</td></tr>
<%}%>
</table>
</body>
```

listAll.jsp 页面首先获取 request 中的 all 对象，接着遍历 all 对象输出每条记录的 name 和年龄。

小　　结

本章主要介绍了在 JSP 中使用的 JavaBean 规范，JSP 页面中使用脚本段代码和使用动作元素访问 JavaBean。实际开发中，对于简单的小型应用可以使用 JSP＋JavaBean 开发模式，对于大型应用采用 JSP＋Servlet＋JavaBean 开发模式或 JSP＋Spring MVC＋JavaBean 开发模式。

思考与习题

1. 请写出 JavaBean 规范。
2. 举例说明使用脚本段代码访问 JavaBean 的步骤。

3. 请写出在用户注册时,使用动作元素＜jsp:useBean＞和＜jsp:setProperty＞的 4 种用法,将用户名和密码设置到 JavaBean 的属性中。

4. 请写出使用 JSP＋JavaBean 开发模式实现用户登录和用户列表功能的关键代码。

5. 请写出使用 JSP＋Servlet＋JavaBean 开发模式实现用户登录和用户信息修改功能的关键代码。

第6章

表达式语言

本章介绍的表达式语言全称是 JSP 表达式语言，又称为 EL(Expression Language)，主要用于获取 JSP 作用域(page、request、session、application)的变量或 JavaBean 对象。表达式语言是 JSP 2.0 规范的一部分，支持 Servlet 2.4/JSP 2.0 以上规范的 Servlet 容器都可以运行包含表达式语言的 JSP 页面。JSP 页面应尽可能使用 JSP 表达式语言替代 JSP 页面脚本段。

6.1 表达式语言基础

6.1.1 表达式语言语法

表达式语言语法格式：

```
${ 表达式 }
```

表达式语言以 ${ 为开始，以 } 为结尾，中间部分是表达式。例如，getUserName.jsp 使用表达式语言获取 request 作用域中的 JavaBean 对象 user 的 name 属性值，其关键代码如下：

```
<jsp:useBean id="user" class="ch06.User" scope="request"/>
<jsp:setProperty name="user" property="name" value="Mike"/>
userName=${user.name}
```

在浏览器地址栏输入 http://localhost:8080/ch06-01/getUserName.jsp 后按 Enter 键，显示结果如图 6-1 所示。

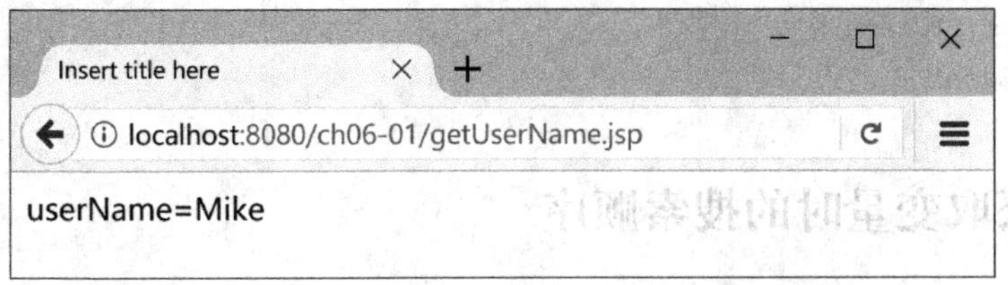

图 6-1 使用 EL 获取 JavaBean 对象 user 的 name 属性

此处使用了.运算符获取 JavaBean user 的 name 属性值。在表达式语言中还可以使用[]运算符。

6.1.2 .运算符与[]运算符

表达式语言提供了.和[]运算符来获取某个 JSP 作用域的数据。在通常情况下两者的效果是一样的。

例如,如下 getUserName2.jsp 演示了使用.和[]运算符获取 JavaBean user 的 name 属性。

```
<jsp:useBean id="user" class="ch07.User" scope="request"/>
<jsp:setProperty name="user" property="name" value="Mike"/>
userName=${user.name}
userName=${user["name"]}
```

还可以同时混合使用.和[]运算符。例如,如下代码同时混合使用.和[]运算符:

```
<%
    User[] users=new User[3];
    for(int i=0;i<3;i++){
        User u=new User();
        u.setName("user-"+i);
        users[i]=u;
    }
    request.setAttribute("users",users);
%>
${users[0].name}}
```

以下两种情况只能使用[]运算符,不能使用.运算符。

(1) 待获取的属性名称中含有.或-等非字母或非数字的特殊符号。

例如,user 的 name 属性如果改为 na_me,就不能使用 userName=${user.na_me},而应该使用 userName=${user["na_me"]}。

(2) 待获取的 JavaBean 由一个变量指定。

例如,在如下代码中,i 是一个变量,不能使用 ${users.i.name },而只能使用 ${users[i].name }。

```
<c:forEach var="i" begin="0" end="3">
  ${users[i].name }
</c:forEach>
```

6.1.3 获取变量时的搜索顺序

在表达式语言中,JSP 的 4 个作用域 page、request、session、application 分别用

pageScope、requestScope、sessionScope、applicationScope 表示。

1. 获取某个特定作用域中的变量

语法格式：

```
${certainScope.varName}
```

certainScope 是 pageScope、requestScope、sessionScope、applicationScope 之一。如果在 certainScope 指定的作用域找到 varName，则返回 varName 的值；如果在 certainScope 指定的作用域没有找到 varName，则返回 null。

例如，如下代码展示了表达式语言获取 4 个 JSP 作用域中的变量 userName

```
<%
    pageContext.setAttribute("userName","page",pageContext.PAGE_SCOPE);
    request.setAttribute("userName","request");
    session.setAttribute("userName","session");
    application.setAttribute("userName","application");
%>
page`userName:${pageScope.userName }<br/>
request`userName:${requestScope.userName }<br/>
session`userName:${sessionScope.userName }<br/>
application`userName:${applicationScope.userName }<br/>
```

2. 获取变量时不指定作用域

语法格式：

```
${varName}
```

当不指定作用域时，表达式语言获取变量会按照 page、request、session、application 的顺序来搜索变量，即先搜索 page 作用域是否存在 varName，如果存在，则返回该变量的值；如果不存在，再搜索 request 作用域，以此类推，最后搜索 application 作用域，如果在这 4 个作用域都没有找到 varName，则返回 null。

例如，如下代码展示了不指定作用域时表达式语言按照顺序搜索 JSP 作用域变量：

```
<%
    pageContext.setAttribute("userName","page",PageContext.PAGE_SCOPE);
    request.setAttribute("userName","request");
    session.setAttribute("userName","session");
    application.setAttribute("userName","application");
%>
    page`userName:${userName }
```

每个作用域都有 userName，表达式语言首先搜索到 page 作用域中的 userName。页面显示的结果如下：

```
page`userName:page
```

6.1.4 自动转型

表达式语言除了提供获取变量的语法外，还有另外一个功能是自动转型。

例如，${param.age+5}表示获得请求参数 age 的值后与整数 5 相加。

JSP 页面传递的参数都是字符类型，在 Java 中对应的是 String。假如页面传来的 age 是 20，那么上面的结果是 25。${param.age+5}相当于在 JSP 或 Servlet 中的如下 Java 代码：

```
String sage=request.getParameter("age");
int age=Integer.parseInt(sage);
age=age+5;
```

即表达式语言将 String 的参数首先变成了 int 类型的 20 后，然后与 5 相加。

6.1.5 保留字

保留字是系统使用的标识符，不允许用户在程序中使用这些标识符。表达式语言保留字见表 6-1。

表 6-1 表达式语言保留字

and	eq	gt	true	or	ne	le	false
no	lt	ge	null	instanceof	empty	div	mod

6.1.6 内建对象

在第 4 章介绍了 JSP 内建对象。表达式语言也有自己的内建对象。表达式语言的内建对象说明见表 6-2。

表 6-2 表达式语言的内建对象说明

内建对象	类　型	说　明
pageContext	javax.servlet.ServletContext	JSP 页面上下文
pageScope	java.util.Map	页面作用域
requestScope	java.util.Map	请求作用域
sessionScope	java.util.Map	会话作用域
applicationScope	java.util.Map	应用作用域

续表

内建对象	类　型	说　明
param	java. util. Map	表示获取页面传来的某个请求参数的值。param. parameterName 相当于 request. getParameter("paramName")
paramValues	java. util. Map	表示获取页面传来的请求参数的所有值。paramValues. paramName 相当于 request. getParameterValues("paramName")
header	java. util. Map	表示获取客户端提交请求中的 header。header. headerName 相当于 request. getHeader("headerName")
headerValues	java. util. Map	表示获取客户端提交请求中的 header 中的所有值
cookie	java. util. Map	表示获取客户端的 cookie
initParam	java. util. Map	表示获取 JSP 页面在 web. xml 中的初始化参数

表达式语言内建对象的含义与 JSP 内建对象相同,但是仅仅是用来获取参数值,不如 JSP 内建对象功能强大。例如,在 session 作用域中存储了一个变量,它的名称为 username,在 JSP 中使用 session. getAttribute("username") 来取得 username 的值;但在表达式语言中,则是使用 ${sessionScope. username}来取得其值的。

6.1.7 运算符

1. 算术运算符

表达式语言的算术运算符主要有 5 个,其使用说明见表 6-3。

表 6-3　表达式语言的算术运算符使用说明

算术运算符	说　明	范　例	结　果
+	加	${15 + 3}	18
-	减	${15 - 3}	12
*	乘	${15 * 3}	45
/或 div	除	${15 / 3}	5
%或 mod	取余	${15 % 3}	0

2. 关系运算符

表达式语言关系运算符有 6 个，其使用说明见表 6-4。

表 6-4　表达式语言的关系运算符使用说明

关系运算符	说　明	范　例	结　果
＝＝或 eq	判断是否相等	${8 ＝＝ 8}或 ${8 eq 8}	true
！＝或 ne	判断是否不等于	${8 ！＝ 8}或 ${8 ne 8}	false
＜或 lt	判断是否小于	${5 ＜ 8} 或 ${5 lt 8}	true
＞或 gt	判断是否大于	${5 ＞ 8} 或 ${5 gt 8}	false
＜＝或 le	判断是否小于或等于	${5 ＜＝ 8} 或 ${5 le 8}	true
＞＝或 ge	判断是否大于或等于	${5 ＞＝ 8} 或 ${5 ge 8}	false

3. 逻辑运算符

表达式语言逻辑运算符共有 3 个，其使用说明见表 6-5。

表 6-5　表达式语言的逻辑运算符使用说明

<table>
<tr><th>逻辑运算符</th><th>说　明</th><th>范　例</th><th>结　果</th></tr>
<tr><td>&& 或 and</td><td>与</td><td>${A && B}或 ${A and B}</td><td>true / false</td></tr>
<tr><td>||或 or</td><td>或</td><td>${A || B}或 ${A or B}</td><td>true / false</td></tr>
<tr><td>！或 not</td><td>非</td><td>${ ！A }或 ${not A}</td><td>true / false</td></tr>
</table>

4. 其他运算符

表达式语言除了前面介绍的 3 大类运算符外，还有几个重要的运算符，其使用说明见表 6-6。

表 6-6　表达式语言的其他运算符

其他运算符	说　明	范　例	结　果
empty	判断是否为 null 或空	${empty requestScope.userName}	true /false
？：	条件运算符	${ A ？B ：C}	当 A 为 true 执行 B，否则执行 C
( )	改变优先权	${ A * (B+C)}	A 乘以 B 和 C 的和

5. 运算符优先级

运算符优先级见表 6-7，优先级顺序是由高至低、由左至右。

表 6-7 运算符优先级

优先级	运算符
高	[]、.
↓	()
	－(负)、not(或!)、empty
	*、/(或 div)、%(或 mod)
	＋、－(减)
	<、>、<=、>=、lt、gt、le、ge
	==、!=、eq、ne
	&&、and
	||、or
低	?　:

6.2 表达式语言函数

表达式语言允许定义函数和JSP页面使用表达式语言调用函数。

6.2.1 表达式语言定义函数

表达式语言定义一个函数的步骤如下：编写一个 Java 类，该类的方法必须是 static 和 public 的；创建标签库描述文件，在该文件中描述函数与对应的 Java 类的映射；在 web.xml 中描述函数。

下面定义一个能实现两个数字相加的函数。

1. 编写 Java 类

```
package ch06;
public class AddFunction {
    public static int add(int x ,int y){
        return x+y;
    }
}
```

2. 创建标签库描述文件 addFunction.tld

在 WEB-INF 下创建 addFunction.tld。

```
<?xml version="1.0" encoding="UTF-8"?>
```

```
<taglib xmlns:javaee="http://java.sun.com/xml/ns/j2ee"
    xmlns:xsi="http://www.w3.org/2001/XMLSchema-instance"
    xsi:schemaLocation=
      "http://java.sun.com/xml/ns/j2ee/web-jsptaglibrary_2_0.xsd" version="2.0">
    <tlib-version>1.0</tlib-version>
    <jsp-version>1.2</jsp-version>
    <short-name>myfun</short-name>
    <uri>http://www.sdjzu.edu.cn/function</uri>
    <description>add function</description>
    <function>
        <name>add</name>
        <function-class>ch06.AddFunction</function-class>
        <function-signature>int add(int,int)</function-signature>
    </function>
</taglib>
```

- <taglib>是根元素。
- <short-name>指定该标签的简写名称。
- <uri>用于指定该标签库的网络位置。
- <function>元素用于描述表达式语言函数与对应的 Java 类的映射。
- <name>子元素指定函数名称。
- <function-class>子元素指定函数对应的 Java 类的全名(包名.类名)。
- <function-signature>指定函数对应的方法定义。

3. 在 web.xml 中描述函数

```
<jsp-config>
    <taglib>
        <taglib-uri>http://www.sdjzu.edu.cn/function</taglib-uri>
        <taglib-location>/WEB-INF/addFunction.tld</taglib-location>
    </taglib>
</jsp-config>
```

6.2.2 JSP 页面使用表达式语言调用函数

调用上述函数的 JSP 页面 el_function.jsp 的关键代码如下：

```
<%@page language="java" contentType="text/HTML; charset=UTF-8"
    pageEncoding="UTF-8"%>
<%@taglib prefix="myfun" uri="http://www.sdjzu.edu.cn/function" %>
<html><head><title>el_function.jsp</title></head>
<body>
请输入 2 个数字,点击 =号后显示结果<br/>
```

```
<form action="el_function.jsp">
    <input type="text" name="firstNum" value="${param.firstNum}"/>
      +
    <input type="text" name="secondNum"
      value="${param.secondNum }"/>
    <input type="submit" value="=">
    <input type="text" name="sum"
      value="${myfun:add(param.firstNum,param.secondNum)}"/>
</form>
</body>
</html>
<%@taglib prefix="myfun" uri="http://www.sdjzu.edu.cn/function" %>
```

使用标签库指令指定标签库的前缀为 myfun,uri 为 http://www.sdjzu.edu.cn/function。

<form>表单中的前面的 input 标记用于让用户输入 2 个数字,后面的 input 标记为提交,使用=作为提交按钮的显示。

下面的 input 标记使用表达式语言调用函数,接收用户输入的 2 个数字并求和。

```
<input type="text" name="sum"
    value="${myfun:add(param.firstNum,param.secondNum)}"/>
```

在输入 3 和 5 后,el_function.jsp 执行结果如图 6-2 所示。

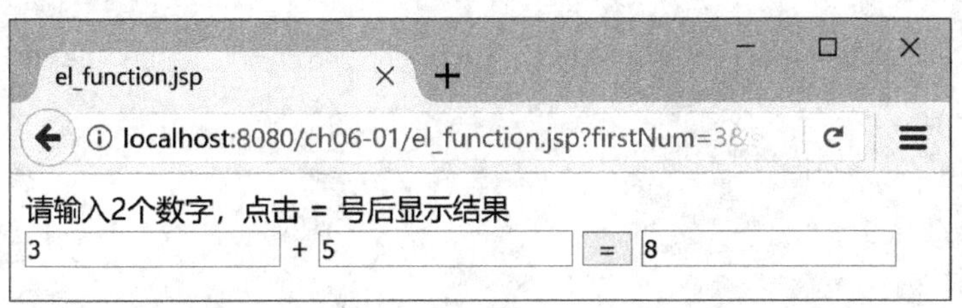

图 6-2　输入 3 和 5 后 el_function.jsp 执行结果

小　　结

本章主要介绍 JSP 表达式语言的语法知识,包括.和[]运算符、搜索变量时的顺序、表达式语言的自动转型等,以及 JSP 表达式语言中定义函数和在 JSP 页面调用函数。

思考与习题

1. JSP 表达式语言的主要作用是什么?
2. 举例说明.和[]运算符的区别。

3. JSP 的 4 个作用域是 page、request、session、application。表达式语言中用什么表示这 4 个作用域？

4. 获取变量时不指定作用域时表达式语言的搜索顺序是什么？

5. 举例说明表达式语言的自动转型功能。

6. 请写 JSP 表达式语言定义函数的步骤。

7. 请写出使用 JSP 表达式语言定义一个实现乘法的函数，并在 JSP 页面调用该函数的关键代码。

第7章

JSTL

JSTL(Java Server Pages Standard Tag Library)是由JCP(Java Community Process)制定的统一的JSP官方标准标签库。JSTL实现了JSP应用开发中的常见操作,用于替代JSP中的脚本段。本章基于JSTL 1.2版本介绍JSTL标签库。

7.1 JSTL简介

7.1.1 JSTL构成

JSTL由5个子标签库构成:核心标签库、SQL标签库、XML标签库、函数标签库、I18N格式标签库。JSTL的5个子标签库说明见表7-1。

表7-1 JSTL的5个子标签库说明

名称	prefix	uri	使用案例
核心标签库	c	http://java.sun.com/jsp/jstl/core	<c:out>
SQL标签库	sql	http://java.sun.com/jsp/jstl/sql	<sql:query>
XML标签库	xml	http://java.sun.com/jsp/jstl/xml	<x:parse>
函数标签库	fn	http://java.sun.com/jsp/jstl/functions	<fn:split>
I18N格式标签库	fmt	http://java.sun.com/jsp/jstl/fmt	<fmt:formatDate>

7.1.2 在JSP页面使用JSTL

1. 使用taglib指令声明标签库

语法格式:

```
<%@taglib uri="uri" prefix="prefix"%>
```

其中,uri指定该标签库的位置;prefix指定JSP页面中标签库的前缀名称。

2. JSP页面使用标签库中的标签

语法格式:

<前缀名称:标签>

例如,如下 usejstl.jsp 使用 JSTL 核心标签库的 out 标签输出 JavaBean 对象 user 的属性值。

```
<%@page pageEncoding="UTF-8"%>
<%@taglib uri="http://java.sun.com/jsp/jstl/core" prefix="c"%>
<html><head><title></title></head>
<body>
    <jsp:useBean id="user" class="ch07.User"/>
    <jsp:setProperty property="name" name="user" value="Alice"/>
    使用动作元素获取 name 为:
    <jsp:getProperty property="name" name="user"/><br/>
    使用 JSTL 标签获取 name 为:<c:out value="${user.name}"/>
</body>
</html>
```

在<%@ taglib uri="http://java.sun.com/jsp/jstl/core" prefix="c"%>中,uri 指定核心标签库的位置为 http://java.sun.com/jsp/jstl/core;prefix 指定核心标签库在当前 JSP 页面的前缀名称为 c。

<c:out value="${user.name}"/>使用核心标签库的 out 标签输出 user 对象的 name 属性。usejstl.jsp 的运行效果如图 7-1 所示。

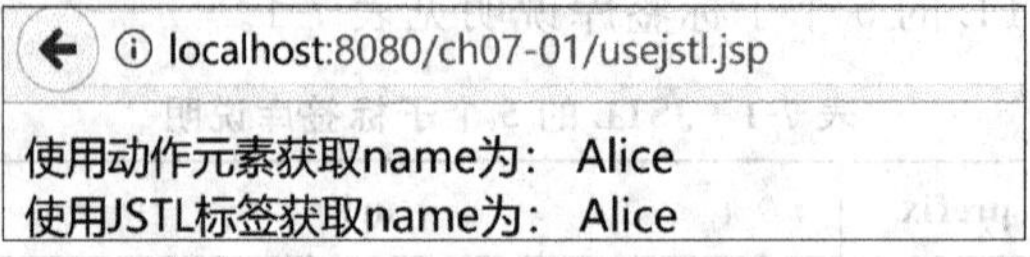

图 7-1 usejstl.jsp 的运行结果

7.2 核心标签库

核心标记库中的标签可以分为 4 类:一般操作、流程控制操作、迭代操作、URL 操作。核心标签库分类见表 7-2。

表 7-2 核心标签库分类

分 类	功能分类	标签名称
核心标签库	一般操作	out、set、remove、catch
	流程控制操作	if、choose、when、otherwise
	迭代操作	forEach、forTokens
	URL 操作	import、url、redirect、param

JSP 中使用核心标签库时，<%@ taglib>指令通常如下所示：

```
<%@taglib uri="http://java.sun.com/jsp/jstl/core" prefix="c"%>
```

prefix="c"指定标签库的前缀名称为 c。可以将 c 写为其他名称。通常为了程序可读性，一般记作 c。

7.2.1 一般操作

1. <c:out>

用于在 JSP 页面输出数据，类似于 JSP 中的表达式<%= Expression>，但是<c:out>功能更为强大。

语法格式 1：无 body。

```
<c:out value="value" [escapeXml="{true|false}"] [default="defaultValue"] />
```

语法格式 2：有 body。

```
<c:out value="value"  [escapeXml="{true|false}"] >
    default value
</c:out>
```

语法格式 2 带有 body，表示如果 value 指定的值为 null，那么将 body 作为默认值输出。

<c:out>标签的属性说明见表 7-3。

表 7-3 <c:out>标签的属性说明

属性	是否必需	属性说明	默认值
value	否	指定输出数据，取值为表达式或常量	无
escapeXml	否	指定是否将特殊字符转换为实体代码，取值为 true 或 false。true 表示将特殊字符如"<"等转换成实体代码；false 表示不将特殊字符转换成实体代码	true
default	否	指定 value 为 null 时输出的默认值，取值为表达式或常量	true

常见特殊字符对应的实体代码见表 7-4。

表 7-4 常见特殊字符对应的实体代码

特殊字符	实体代码
<	<
>	>
&	&
‘	'
“	"

<c:out>典型使用说明见表 7-5。

表 7-5 <c:out>典型使用说明

<c:out>典型使用	说明
<c:out value="${user.name}"/>	在 JSP 页面输出 value 指定的 user 对象的 name 属性值
<c:out value="${user.name}" default="Mike"/>	在 JSP 页面输出 value 指定的 user 对象的 name 属性值。如果 name 不为 null,则在 JSP 页面输出 name 属性值;否则在 JSP 页面输出 default 指定的值 Mike
<c:out value="<p>这是一段文章。</p>" escapeXml="false" />	使用 escapeXml="false"设置将输出数据的特殊字符不转换为实体代码
<c:out value="${user.name}"> Mike </c:out>	在 JSP 页面输出 value 指定的 user 对象的 name 属性值。如果 name 不为 null,则在 JSP 页面输出 name 属性值;否则,在 JSP 页面会将 body 的值 Mike 作为默认值输出

2. <c:set>

用于设置一个变量的值或一个 JavaBean 对象属性的值,并将变量或 JavaBean 对象保存在某个 JSP 作用域中。

语法格式 1:无 body。

```
<c:set var="varName" value="somevalue" [scope="{page|request|session|
application}"] />
```

语法格式 2:有 body。

```
<c:set var="varName" [scope="{page|request|session|application}"] >
    body content
</c:set>
```

语法格式 3:设置 JavaBean 对象属性的值,无 body。

```
<c:set target="target"property="propertyName"value="value"
    [scope="{page|request|session|application}"]/>
```

语法格式 4:设置 JavaBean 对象属性的值,有 body。

```
<c:set target="target"property="propertyName"
    [scope="{page|request|session|application}"]/>
    body content
</c:set>
```

<c:set>标签的属性说明见表 7-6。

<c:set>典型使用说明见表 7-7。

表 7-6　<c:set>标签的属性说明

属性	是否必需	属性说明	默认值
value	否	指定待设置的变量(或 JavaBean 对象属性)的值,取值为一个表达式或常量	无
var	否	指定一个变量来保存 value 的结果	无
scope	否	指定 var 变量的 JSP 作用域	page
target	否	指定一个 JavaBean 或 java. util. Map 对象	无
property	否	指定 target 对象的某个属性名或 Map 的 key	无

表 7-7　<c:set>典型使用说明

<c:set>典型使用	说　明
<c:set var="num6" value="${1+5}" scope="request"/>	设置变量 num6 的值为表达式 ${1+5}的结果,并将 num6 保存在 request 作用域
<c:set var="num7" scope="session" > ${1+6} </c:set>	设置变量 num7 的值为 body 内容,即 ${1+6}的结果,并将 num7 保存在 session 作用域
<jsp:useBean id="user" class="ch07. User"/> <c:set target="${user}" property="age" value="22"/>	设置默认作用域 page 中的 JavaBean 对象 user 的 age 属性值为 22,并将 value 指定的字符串 22 自动转型为整型的 22
<c:set target="${user}" property="age" > 23 </c:set>	设置默认作用域中的 JavaBean 对象 user 的 age 属性值为 body 的内容 23

3. <c:remove>

<c:remove>用于删除某个 JSP 作用域中的变量或 JavaBean 对象。

语法格式:

```
<c:remove var="varName" [scope="{page|request|session|application}"] />
```

<c:remove>标签的属性说明见表 7-8。

表 7-8　<c:remove>标签的属性说明

属性	是否必需	属性说明	默认值
scope	否	指定待删除的变量或 JavaBean 对象的作用域	page
var	是	指定待删除的变量名或对象名	无

<c: remove>典型使用说明见表 7-9。

表 7-9 <c:remove>典型使用说明

<c:remove>典型使用	说　明
<c:remove var="num"scope="request"/>	删除 request 作用域中 num 变量
<c:remove var="num" />	删除某个作用域中的 num 变量。没有指定 scope，<c:remove>将按照 page、request、session、application 的顺序来查找 num 变量。如果在 page 作用域中找到 num 后会删除该变量，否则再到 request 作用域中查找 num，以此类推。如果最后找不到 num 变量，则不执行任何操作
<jsp:useBean id="user" class="ch07.User"/> <c:set target="${user}" property="name" value="Alice"/> <c:remove var="user" scope="page"/>	删除 page 作用域中的 user 对象

4. <c:catch>

<c:catch>用于捕获该标签中脚本段或其他标签抛出的异常，并将异常信息保存起来。

语法格式：

```
<c:catch [var="varName"]>
    可能会发生异常的代码或标签
</c:catch>
```

<c:catch>标签的属性说明见表 7-10。

表 7-10　<c:catch>标签的属性说明

属性	是否必需	属性说明	默认值
var	否	指定一个变量保存异常信息	无

例如，c_catch.jsp 使用<c:catch>标签捕获异常，其关键代码如下：

```
<c:catch var="error">
    <jsp:useBean id="user" class="ch07.User"/>
    <c:set target="${user}" property="age" value="Alice"/>
</c:catch>
${error}
```

在<c:catch>和</c:catch>之间创建了 user 对象，并试图为 user 对象的 age 属性赋值 Alice。user 对象的 age 属性是 int 类型，故将 String 类型的 Alice 赋值给 age 肯定会出错。当发生错误时，<c:catch>捕获异常并将错误信息保存在 error 变量中。最后使用表达式语言输出错误信息 error。c_catch.jsp 的运行结果如图 7-2 所示。

如果删除 c_catch.jsp 中的<c:catch>，会输出抛出异常的堆栈信息，显得页面凌乱。删除<c:catch>后，c_catch.jsp 的运行结果如图 7-3 所示。

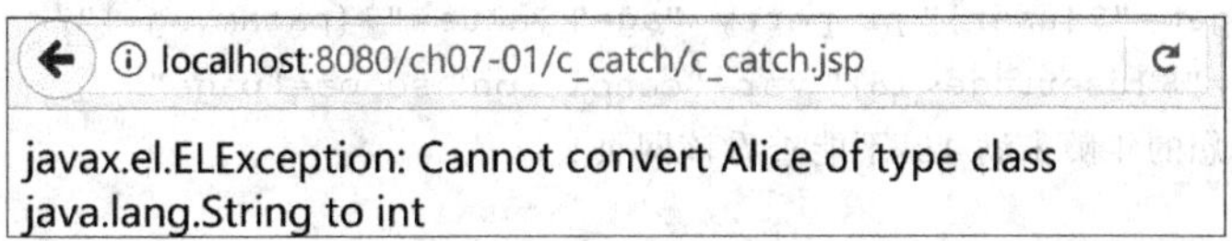

图 7-2 c_catch.jsp 的运行结果

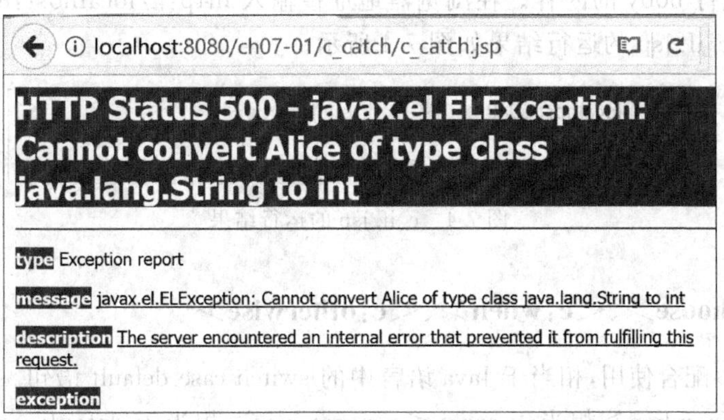

图 7-3 删除<c:catch>后 c_catch.jsp 的运行结果

7.2.2 流程控制操作

1. <c:if>

<c:if>标签用于条件判断，相当于 Java 语言中的 if 语句。

语法格式：

```
<c:if test="testCondition" var="varName" [scope="{page|request|session|
application}"]>
    body content
</c:if>
```

<c:if>的属性说明见表 7-11。

表 7-11 <c:if>的属性说明

属性	是否必需	属性说明	默认值
test	是	指定一个结果为 true 或 false 的条件表达式	无
var	否	指定一个变量保存 test 属性运算后的结果	无
scope	否	指定 var 的作用域	page

如果 test 指定的表达式 testCondition 结果为 true，则执行 body；如果是 false，则不执行 body。

例如，使用<c:if>的 c_if.jsp，其关键代码如下：

```
<jsp:useBean id="user" class="ch07.User" />
```

```
<c:set target="${user}" property="age" value="${param.age}"/>
<c:if test="${user.age<18}" var="condition" scope="page">
    对不起,您的年龄未满 18,不能查看该网页!
</c:if>
```

第三行使用<c:if>判断用户的年龄即 user.age 是否小于 18,如果是则执行 body 的内容;否则不执行 body 的内容。在浏览器地址栏输入 http://localhost:8080/ch07/c_if.jsp? age=17,c_if.jsp 的运行结果如图 7-4 所示。

图 7-4　c_if.jsp 的运行结果

2. <c:choose>、<c:when>、<c:otherwise>

这几个标签配合使用,相当于 Java 语言中的 switch case default 语句。<c:choose>相当于 switch,<c:when>相当于 case,<c:otherwise>相当于 default。<c:choose>的 body 只能是空,或是 0 个或多个<c:when>加上 1 个<c:otherwise>,且<c:when>必须出现在<c:otherwise>之前。

语法格式:

```
<c:choose>
    <c:when test="testCondition1">
        当 testCondition1 为 true 时,执行此段代码
    </c:when>
    <c:when test="testCondition2">
        当 testCondition2 为 true 时,执行此段代码
    </c:when>
    …(其他<c:when>标记)
    <c:otherwise>
        上面所有的<c:when>均不满足条件时,执行此处代码
    </c:otherwise>
</c:choose>
```

<c:when>的属性说明见表 7-12。

表 7-12　<c:when>的属性说明

属性	是否必需	属性说明	默认值
test	是	指定结果是 boolean 的表达式	无

例如,使用<c:choose>的 c_choose.jsp,其关键代码如下:

```
<jsp:useBean id="user" class="ch07.User" />
<c:set target="${user}" property="age" value="${param.age}"/>
```

```
<c:choose>
    <c:when test="${user.age<18}">
        您的年龄未满 18!
    </c:when>
    <c:when test="${user.age>18}">
        您的年龄过了 18!
    </c:when>
    <c:otherwise>
        您的年龄正好 18!
    </c:otherwise>
</c:choose>
```

在浏览器地址栏输入 http://localhost:8080/ch07/c_choose/c_choose.jsp? age=18 后 c_choose.jsp 的运行结果如图 7-5 所示。

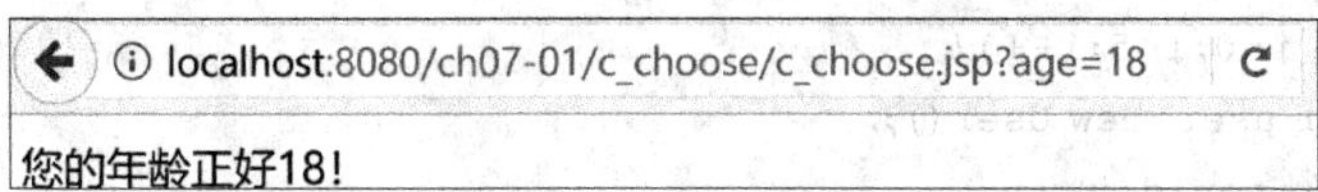

图 7-5　c_choose.jsp 的运行结果

7.2.3　迭代操作

JSTL 中的迭代标签有<c:forEach>和<c:forTockens>。

1. <c:forEach>

<c:forEach>用于迭代操作。

语法格式 1：迭代集合对象的成员。

```
<c:forEach [var="varName"] items="collection" [varStatus="varStatus"] [begin=
"begin"]
    [end="end"] [step="step"]
    body content
</c:forEach>
```

语法格式 2：不浏览集合对象，只是迭代指定的次数。

```
<c:forEach [var="varName"] [varStatus="varStatus"] begin="begin"
    end="end" [step="step"]
    body content
</c:forEach>
```

<c:forEach>的属性说明见表 7-13。

表 7-13　<c:forEach>的属性说明

属性	是否必需	属性说明	默认值
var	否	指定一个变量,保存当前正在迭代的元素	无
items	否	指定迭代的集合对象	无
varStatus	否	存放迭代的状态,可以访问迭代自身的信息	无
begin	否	指定迭代开始的索引	0(表示第一个元素)
end	否	指定迭代结束的索引	迭代结束的索引
step	否	指定迭代的步长	1

(1) 使用<c:forEach>语法格式 1 迭代集合对象。c_forEach_1_1.jsp 的关键代码如下:

```
<%
    Collection all=new ArrayList();
    for(int i=0;i<5;i++){
        User user=new User();
        user.setId(i);
        user.setName("user"+i);
        all.add(user);
    }
    request.setAttribute("all",all);
%>
<c:forEach var="user" items="${all}">
    user.id=${user.id}--->user.name=${user.name} <br/>
</c:forEach>
```

以上代码在<c:forEach>中对 items 指定的从 request 作用域取出的 all 对象进行迭代,var 指定的 user 表示当前正在迭代的 User 对象;在<c:forEach>的 body 中,输出正在迭代的 user 的 id 和 name。c_forEach_1_1.jsp 的运行结果如图 7-6 所示。

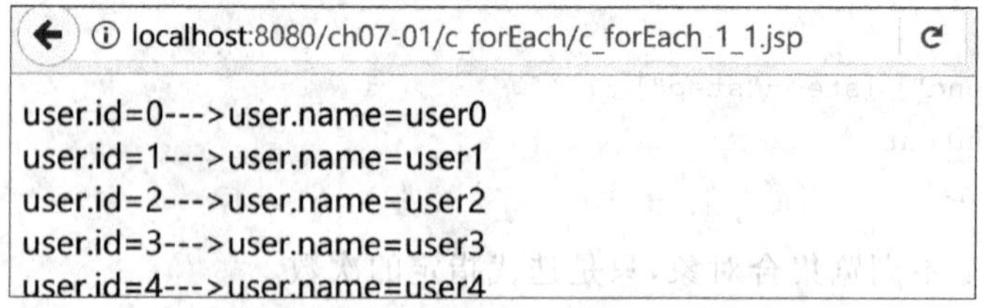

图 7-6　c_forEach_1_1.jsp 的运行结果

(2) <c:forEach>的 varStatus 属性用于存放迭代的状态,可以访问迭代自身的信息。varStatus 的属性说明见表 7-14。

表 7-14　varStatus 的属性说明

属性名	类　型	含　义
index	int	当前迭代元素是集合的第几个元素

续表

属性名	类　　型	含　　义
count	int	当前迭代是第几次迭代
first	boolean	当前迭代是否是第一次迭代
last	boolean	当前迭代是否是最后一次迭代

c_forEach_1_2.jsp 演示了访问迭代自身的信息，其关键代码如下：

```
<%
    Collection all=new ArrayList();
    for(int i=0;i<5;i++){
        User user=new User();
        user.setId(i);
        user.setName("user"+i);
        all.add(user);
    }
    request.setAttribute("all",all);
%>
<c:forEach var="user" items="${all}" begin="1" end="3" step="2" varStatus="s">
    <p>
        user.id=${user.id}--->user.name=${user.name} <br/>
        当前迭代的元素的是集合的第${s.index }个元素<br/>
        当前迭代是第${s.count }次迭代<br/>
        当前迭代是否是第一次迭代:${s.first }<br/>
        当前迭代是否是最后一次迭代:${s.last }<br/>
    </p>
</c:forEach>
```

上述 c_forEach_1_2.jsp 的<c:forEach>中，begin="1"表示从第 2 个元素开始迭代，end="3"表示到第 4 个元素迭代结束，step="2"表示迭代的步长为 2，varStatus="s"表示每次迭代的状态存放在 s 中。在<c:forEach>的 body 中，使用 ${s.index }、${s.count }、${s.first }、${s.last }分别输出每次迭代的状态信息。c_forEach_1_2.jsp 的运行结果如图 7-7 所示。

(3) 使用<c:forEach>语法格式 2 不浏览集合对象，只是迭代指定的次数。c_forEach_2.jsp 的关键代码如下：

```
<c:forEach begin="1" end="3" step="1" var="i">
    ${i} <br/>
</c:forEach>
```

c_forEach_2.jsp 的运行结果如图 7-8 所示。

2. <c:forTokens>

<c:forTokens>按照指定的一个或多个分隔符对字符串进行迭代。

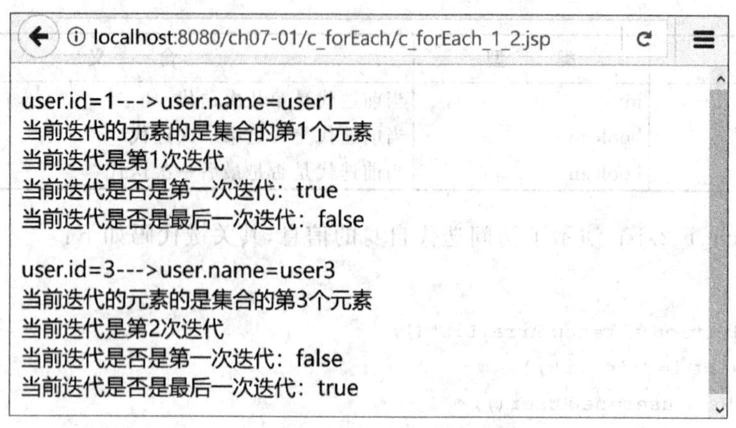

图 7-7　c_forEach_1_2.jsp 的运行结果

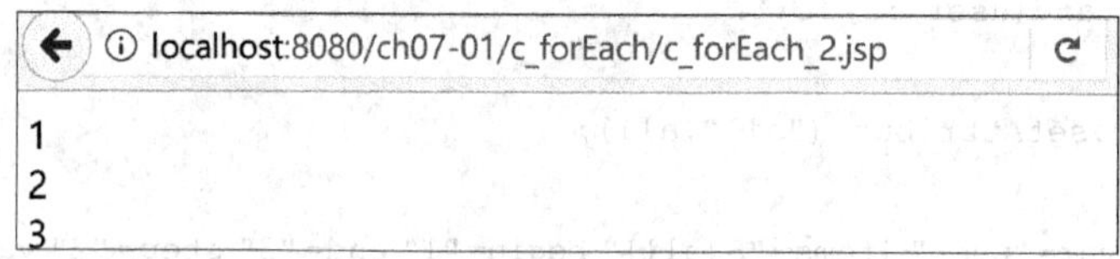

图 7-8　c_forEach_2.jsp 的运行结果

语法格式：

```
<c:forTokens [var="varName"] items="StringOfTokens"delims="delimiters"
   [varStatus="varStatus"] [begin="begin"] [end="end"] [step="step"]
   body content
</c: forTokens>
```

<c:forTokens>标签的主要属性说明见表 7-15。

表 7-15　<c:forTokens>标签的主要属性说明

属性	是否必需	属性说明	默认值
items	否	指定带有分隔符的字符串	无
delims	是	指定字符串分隔符	无

<c:forTokens>标签的属性除了 items 和 delims 外，其他属性如 var、varStatus、begin、end、step 的含义与<c:forEach>一样。items 属性必须是字符串；delims 属性指定字符串分隔符，分隔符将字符串分割为多个子字符串。

例如，使用<c:forTokens>的 c_forTokens.jsp 的关键代码如下：

```
<c:forTokens var="item" items="zhangsan,0531-12345678" delims=",-">
   ${item}
</c:forTokens>
```

delims 属性指定","和"-"作为分隔符，items 属性指定了待分割的字符串"zhangsan,0531-12345678"，使用<c:forTokens>会将字符串按照分隔符分割为 3 个元素：

zhangsan、0531、12345678。在<c:forTokens>的 body 中使用表达式语言进行输出。c_forTokens.jsp 的运行结果如图 7-9 所示。

图 7-9 c_forTokens.jsp 的运行结果

7.2.4 URL 操作

1. <c:import>

<c:import>用于将其他静态文件或动态文件包含进来。<c:import>与<jsp:include>类似，但是比<jsp:include>功能更强大。<c:import>和<jsp:include>的最大差别是<jsp:include>只能包含同一个 Web 应用下的文件；<c:import>不仅能包含同一个 Web 应用下的文件，还能包含其他 Web 应用下的文件。

语法格式：

```
<c:import url="url" [context="context"] [var="varName"]
    [scope="{page|request|session|application}"] [charEncoding="charEncoding"]>
    body content
</c:import>
```

<c:import>的属性说明见表 7-16。

表 7-16 <c:import>的属性说明

属性	是否必需	属性说明	默认值
url	是	指定被包含文件的 URL	无
context	否	指定被包含文件所属的 Web 上下文；包含相同的容器下其他 Web 应用必须以/开头	无
var	否	指定一个变量保存 String 类型的 URL	无
scope	否	指定 var 的 JSP 作用域	无
charEncoding	否	指定被包含文件的字符编码	无

例如，使用<c:import>包含百度首页的 c_import.jsp，其关键代码如下：

```
<c:import url="http://www.baidu.com " charEncoding="UTF-8"></c:import>
```

url="http://www.baidu.com"用于指定被包含的 URL 是百度首页；charEncoding="UTF-8"指定被包含的文件的字符编码是 UTF-8。c_import.jsp 的运行结果如图 7-10 所示。

包含同一个 Web 容器中另一个 Web 应用中的页面，需要使用 context 属性。假设当

图 7-10　c_import.jsp 的运行结果

前服务器 Tomcat 上有另一个名为 others 的 Web 应用，others 应用下有一个文件夹 jsp，里面有 index.html 文件，那么就可以写成如下方式将 index.html 包含进来：

```
<c:import url="/jsp/index.html" context="/others"></c:import>
```

被包含的 Web 站点必须在%Tomcat%/conf/server.xml 中被配置为虚拟目录，且<Context>节点的 crossContext 属性为 true，这样，others 站点的文件才能被其他 Web 站点调用。配置虚拟目录，需要在%Tomcat%/conf/server.xml 中的<host>和</host>之间加入如下代码：

```
<Server>
    <!--其他元素略…-->
        <Context path="/others" docBase="d:\others" debug="0" reloadable="true"
            crossContext="true">
        </Context>
      </Host>
    </Engine>
  </Service>
</Server>
```

2. <c:redirect>

<c:redirect>用于将客户请求重定向到另一个资源。

语法格式 1：重定向到其他 url。

```
<c:redirect url="url" [context="context"] />
```

语法格式 2：重定向到其他 url 并传递参数。

```
<c:redirect url="url" [context="context"] >
    <c:param>
</c:redirect>
```

<c:redirect>的属性说明见表 7-17。

表 7-17　<c:redirect>的属性说明

属性	是否必需	属性说明	默认值
url	是	指定另一个资源的 url	无

续表

属性	是否必需	属性说明	默认值
context	否	指定另一个资源所属的 Web 上下文；重定向到相同的容器下其他 Web 应用必须以/开头	当前应用上下文

例如，从 c_redirect.jsp 重定向到 c_redirect_2.jsp 并传递参数，c_redirect.jsp 的关键代码如下：

```
<c:redirect url="c_redirect_2.jsp">
    <c:param name="userName" value="Mike"/>
    <c:param name="userPassword" value="123"/>
</c:redirect>
```

c_redirect.jsp 的＜c：redirect＞body 中使用＜c：param＞传递参数，参数名是 userName，参数值是 Mike。

c_redirect_2.jsp 的关键代码如下：

```
<body>
    这是从 c_redirect.jsp 重定向到的页面 c_redirect_2.jsp。<br/>
    传递的参数 userName:${param.userName}  userPassword:${param.userPassword}
</body>
```

c_redirect_2.jsp 的运行结果如图 7-11 所示。

localhost:8080/ch07-01/c_redirect/c_redirect_2.jsp?userName=Mike&userPassword=123

这是从c_redirect.jsp重定向到的页面c_redirect_2.jsp。
传递的参数userName：Mike userPassword：123

图 7-11　c_redirect.jsp 的运行结果

3. ＜c：url＞

＜c：url＞标签用于产生一个新的 URL。

语法格式 1：无 body。

```
<c:url value="value" [context="context"] [var="varName"]
    [scope="{page|request|session|application}"] />
```

语法格式 2：有 body。

```
<c:url value="value" [context="context"] [var="varName"]
    [scope="{page|request|session|application}"] >
    <c:param>
</c:url>
```

＜c：url＞标签的属性说明见表 7-18。

表 7-18 <c:url>标签的属性说明

属性	属性类型	是否必需	属性说明	默认值
value	String	是	指定产生的新 URL	无
context	String	否	指定新 URL 所属的 Web 上下文;在相同的容器下其他 Web 应用必须以/开头	无
var	String	否	指定一个变量保存 String 类型的 URL	无
scope	String	否	指定 var 的 JSP 作用域	page

例如,c_url.jsp 演示了在<a>标签中使用<c:url>构造新 URL 及使用<c:param>向新 URL 传递参数。c_url.jsp 的关键代码如下:

```
<a href="<c:url value="http://www.baidu.com">
    <c:param name="name" value="Mike"/></c:url>" >去百度搜索
</a>
```

c_url.jsp 的运行结果如图 7-12 所示。

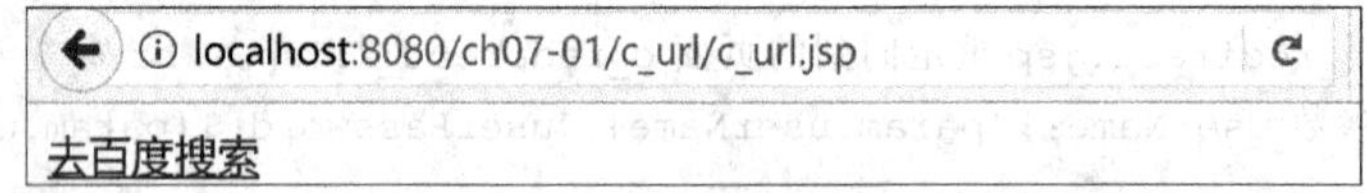

图 7-12 c_url.jsp 的运行结果

单击“去百度搜索”后显示结果如图 7-13 所示,在浏览器地址栏中能看到传递的参数 name 及其值 Mike。

图 7-13 单击去百度搜索后的显示结果

<c:url>的其他 3 个属性 context、var、scope 与<c:import>中的含义相同。

7.3 I18N 格式标签库

有时候一个应用需要支持不同语言和地区的用户访问,这称作应用国际化(Internationalization)。不同语言和地区的用户在访问该应用时,应用能按照该用户所属语言和地区的习惯来显示时间日期、数字以及按照该地区的语言显示字符串消息,这称作应用的本地化(Localization)。例如,中国的用户发出请求时,应用程序应该显示简体中文、按照中国的习惯显示时间日期和数字;当美国的用户发出请求时,显示英文、按照美国的习惯显示时间日期和数字。

国际化有时候称为I18N，这是因为Internationalization这个单词为i～n共有18个字母。一个国际化的应用本地化主要包括消息本地化、时间日期本地化、货币本地化、图像的本地化等。

JSTL国际化部分的标签库被称为I18N格式标签库，能够进行消息本地化、数字时间日期及货币的本地化。I18N格式标签库详细分类分见表7-19。

表7-19 I18N格式标签库详细分类

分类	功能分类	标签名称
I18N标签库	国际化标签	setLocale，requestEncoding
	消息标签	bundle，setBundle，message，param
	数字日期格式化标签	formatNumber，parseNumber，formatDate，parseDate，setTimeZone，timeZone

JSP中使用I18N格式标签库时，<%@ taglib>指令通常如下所示：

```
<%@taglib prefix="fmt" uri="http://java.sun.com/jsp/jstl/fmt" %>
```

7.3.1 国际化标签

1. <fmt:setLocale>

<fmt:setLocale>标签用于设置区域。<fmt:setLocale>标签的属性说明见表7-20。

语法格式：

```
<fmt:setLocale value="locale" [variant="variant"]
   [scope="{page|request|session|application}"] />
```

表7-20 <fmt:setLocale>标签的属性说明

属性	是否必需	属性说明	默认值
value	是	指定某个区域，如zh_CN、zh_TW、en_US	无
variant	否	variant属性可以是供货商或浏览器的规格，比如WIN代表Windows	无
scope	否	表示value属性设置的区域所属的JSP作用域	page

一个区域可以是语言代码或者是语言代码_地区代码。几个常见的区域代码见表7-21。

表7-21 几个常见的区域代码

区域代码	语言(国家/地区)
en	英文
en_US	英文(美国)
zh_CN	中文(中国)

例如，fmt_setLocale.jsp演示了使用<fmt:setLocale>标签设置区域，其关键代码

如下：

```
<%
   Date now =new Date();
   request.setAttribute("now",now);
%>
   <fmt:setLocale value="en_US"/>
   US:<fmt:formatDate value="${now}"/><br>
   <fmt:setLocale value="zh_CN"/>
   China:<fmt:formatDate value="${now}"/><br>
```

在 fmt_setLocale.jsp 中，依次使用<fmt:setLocale>设置区域为英文（美国）、中文（中国），并依次使用格式化日期输出标签<fmt:formatDate>，将日期的显示转化成对应地区的表示习惯。fmt_setLocale.jsp 的执行结果如图 7-14 所示。

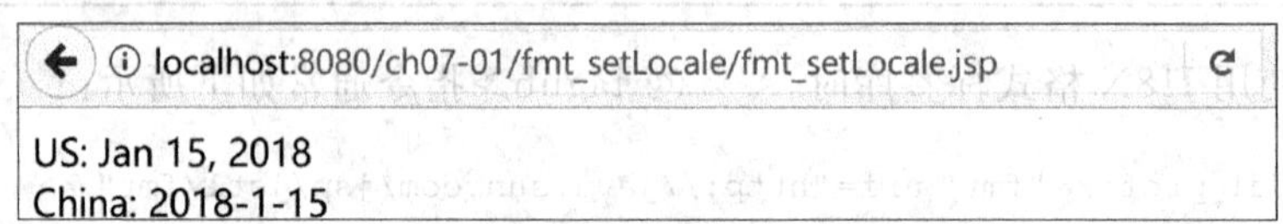

图 7-14　fmt_setLocale.jsp 的执行结果

2. <fmt:requestEncoding>

<fmt:requestEncoding>标签用于设置请求的字符编码格式，功能与 ServletRequest.setCharaterEncoding()相同，能解决 JSP 中 POST 请求的中文乱码问题。

语法格式：

```
<fmt:requestEncoding [value="encodingName"]/>
```

<fmt:requestEncoding>标签的属性说明见表 7-22。

表 7-22　<fmt:requestEncoding>标签的属性说明

属性	类型	是否必需	属性说明	默认值
value	String	否	字符编码类型，如 UTF-8	自动寻找

例如，fmt_requestEncoding_2.jsp 演示了使用<fmt:requestEncoding>设置请求编码为 UTF-8。fmt_requestEncoding.jsp 使用 POST 提交用户名，其关键代码如下：

```
<form action="fmt_requestEncoding_2.jsp" method="post">
   用户名:<input type="text" name="userName">
          <input type="submit" value="提交">
</form>
```

fmt_requestEncoding_2.jsp 接收 fmt_requestEncoding.jsp 提交的用户名参数，并使用了<fmt:requestEncoding>设置请求编码为 UTF-8。fmt_requestEncoding_2.jsp 的关键代码如下：

```
<fmt:requestEncoding value="UTF-8"/>
您的用户名:${param.userName}
```

fmt_requestEncoding.jsp 的运行效果如图 7-15 所示。

图 7-15 fmt_requestEncoding.jsp 的运行效果

在 fmt_requestEncoding.jsp 页面输入“黎明”，单击“提交”按钮后，fmt_requestEncoding_2.jsp 的运行效果如图 7-16 所示。

图 7-16 fmt_requestEncoding_2.jsp 的运行效果

如果<fmt:requestEncoding>标签没有使用 value 属性指定请求的编码类型，例如<fmt:requestEncoding />。这种情况下 Web 容器会首先检测 HTTP 报头 Content-Type 字段中是否包含有 charset 定义，如果有就使用 charset 定义的编码；如果没有定义 charset，那么就在 session 中搜索 javax.servlet.jsp.jstl.fmt.request.charset 变量；如果该变量也不存在，那么就会使用默认字符集 ISO-8859-1 进行解码。

7.3.2 消息标签

JSTL 的消息国际化依赖于 Java SE 的资源包。一个资源包是一组资源文件。资源文件的命名规则是资源包基本名_语言代码_地区代码.properties。每个资源文件中以键值对形式存储对应语言的文本消息，键都是相同的，不同的是文件中每个键的值不同。通常应用还有一个资源包基本名.properties 的默认资源文件，用于当不能找到客户区域对应的资源文件时需要使用该资源文件来显示默认消息。

使用 JSTL 消息标签的 Eclipse 项目 ch07-02 中，会用到 3 个国际化消息的资源文件：Resources_en.properties、Resources_zh.properties 和 Resources.properties，分别存放有关英文消息、中文消息和默认语言消息。Resources.properties 代码与 Resources_en.properties 完全相同。

Resources_en.properties 的代码如下：

```
login.title=User Login
login.userName=User Name
login.password=Password
login.login=Login
```

Resources_zh.properties 代码如下：

```
login.title=\u7528\u6237\u767B\u5F55
login.userName=\u7528\u6237\u540D
login.password=\u5BC6\u7801
login.login=\u767B\u5F55
```

对于 Tomcat 服务器，JSTL 国际化资源文件放置在%Tomcat%\webapps\YourWebApp\WEB-INF\classes 下。在 Eclipse 中的 Java Web 项目 YourWebApp\src 下创建的资源文件会自动部署到%Tomcat\webapps\YourWebApp\WEB-INF\classes 中，所以在 Eclipse Java Web 项目中需要在 src 下创建资源文件。

1. <fmt:bundle>

<fmt:bundle>标签用于设置其标签体中的所有<fmt:message>需要使用的资源包，这样就不必为每个<fmt:message>指定资源包。

语法格式：

```
<fmt:bundle baseName="basename" [prefix="prefix"]>
<fmt:message key="key">
</fmt:bundle>
```

<fmt:bundle>的属性说明如表 7-23 所示。

表 7-23 <fmt:bundle>的属性说明

属性	是否必需	属性说明	默认值
basename	是	指定资源包的基本名	无
prefix	否	指定<fmt:message>标签 key 属性的前缀	无

2. <fmt:setBundle>

<fmt:setBundle>标签用于设置所使用的资源包，或者将资源包保存在 JSP 某个作用域中的变量中。<fmt:setBundle />是没有 body 的标签。

语法格式：

```
<fmt:setBundle basename="basename" [var="varName"]
    [scope="page|request|session|application"] />
```

<fmt:setBundle>标签的属性说明见表 7-24。

表 7-24 <fmt:setBundle>标签的属性说明

属性	是否必需	属性说明	默认值
basename	是	指定资源包的基本名	无
var	否	保存资源包的变量名	无
scope	否	var 的 JSP 作用域	page

3. <fmt:message>

<fmt:message>标签用于查找资源文件中某个键对应的值并进行显示。

语法格式 1：有 key 属性且无 body。

```
<fmt:message key="key" [bundle="resourceBundle"]
    [var="varName"]
    [scope="{page|request|session|application}]"/>
```

语法格式 2：有 key 属性且有 body，body 是<fmt:param>。

```
<fmt:message key="key" [bundle="resourceBundle"]
    [var="varName"]
    [scope="{page|request|session|application}]">
    <fmt:param>
</fmt:message>
```

语法格式 3：没有 key 属性且有 body，body 是 key 和<fmt:param>。

```
<fmt:message [bundle="resourceBundle"] [var="varName"]
    [scope="{page|request|session|application}]"/>
    key
    <fmt:param>
</fmt:message>
```

<fmt:message>标签的属性说明见表 7-25。

表 7-25 <fmt:message>标签的属性说明

属性	是否必需	属性说明	默认值
key	否	指定待获取的文本的键	无
bundle	否	指定资源文件名称	无
var	否	指定一个变量来保存获取的键值	无
scope	否	指定 var 的 JSP 作用域	page

Eclipse 项目 ch07-02 演示了使用消息标签实现用户登录页面的国际化。项目 ch07-02 的结构如图 7-17 所示。

登录页面 login.jsp 的关键代码如下：

```
...
<%@taglib uri="http://java.sun.com/jsp/jstl/fmt" prefix="fmt" %>
<fmt:bundle basename="Resources" prefix="login.">
<html><head><title><fmt:message key="title" /></title></head>
<body>
<form action="" method="post">
    <table>
        <tr><td colspan="2" align="center"><fmt:message key="title"/></td></tr>
```

```
        <tr><td><fmt:message key="userName"/>:</td>
            <td><input type="text" name="name"/></td>
        </tr>
        <tr><td><fmt:message key="password"/>:</td>
            <td><input type="password" name="password"/></td>
        </tr>
        <tr><td colspan="2" align="center">
            <input type="submit" value="<fmt:message key="login"/>"/>
          </td>
        </tr>
    </table>
</form>
</body>
</html>
</fmt:bundle>
```

将该项目部署到 Tomcat 并启动 Tomcat 后，在浏览器中输入 http://localhost:8080/ch07-02/login.jsp，登录界面显示中文如图 7-18 所示。

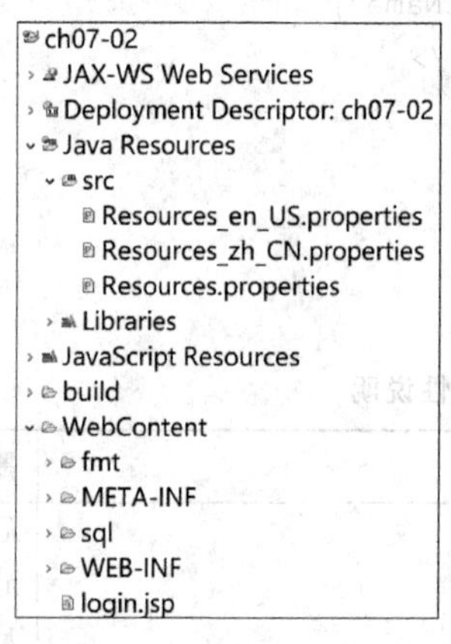

图 7-17　项目 ch07-02 的结构

图 7-18　登录页面显示中文

这是因为在中文版火狐浏览器默认的区域设置(Locale)是中文/中国[zh-cn]。

修改火狐浏览器的 Locale 为英语/美国[en-us]。单击火狐浏览器右上角的打开菜单命令，选择“选项”选项，如图 7-19 所示。

选择选项窗口左侧的内容，找到右侧窗口的语言并单击“选择”按钮，火狐浏览器的 Local 默认设置为“中文/中国[zh-cn]”，如图 7-20 所示。

选择列表中的“英语/美国[en-us]”并单击“上移”按钮将其移动到顶部，表示设置火狐浏览器将首先使用的 Locale 为“英语/美国[en-us]”，如图 7-21 所示。单击“语言和字符编码”窗口的“确定”按钮，再关闭选项窗口。

设置火狐浏览器 Locale 为“英语/美国[en-us]”后，刷新登录界面后的效果如图 7-22 所示。

使用不同的 Locale 同样的登录页面会有不同的结果，JSTL 很容易完成国际化工作。

图 7-19　火狐浏览器打开菜单选择选项

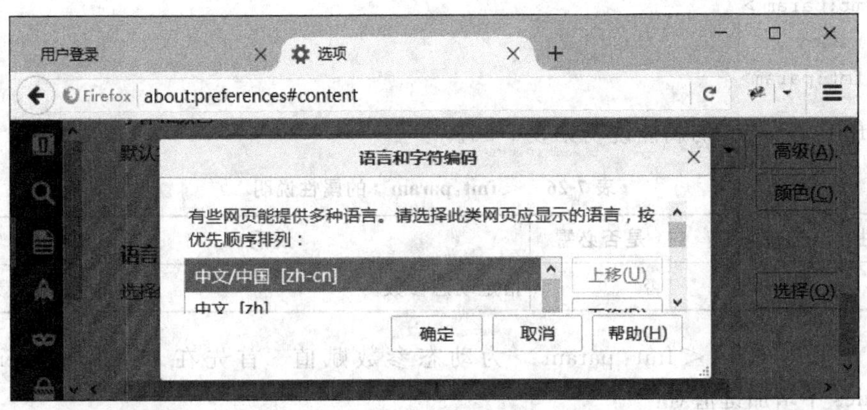

图 7-20　火狐浏览器的 Local 默认设置为"中文/中国[zh-cn]"

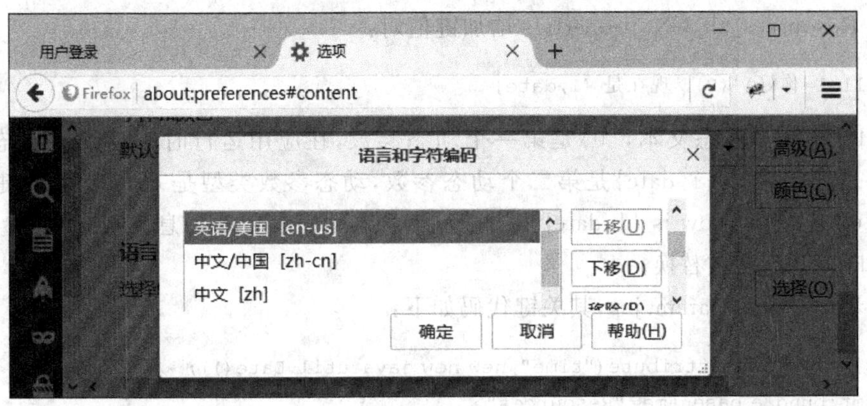

图 7-21　设置火狐浏览器首先使用的 Locale 为"英语/美国[en-us]"

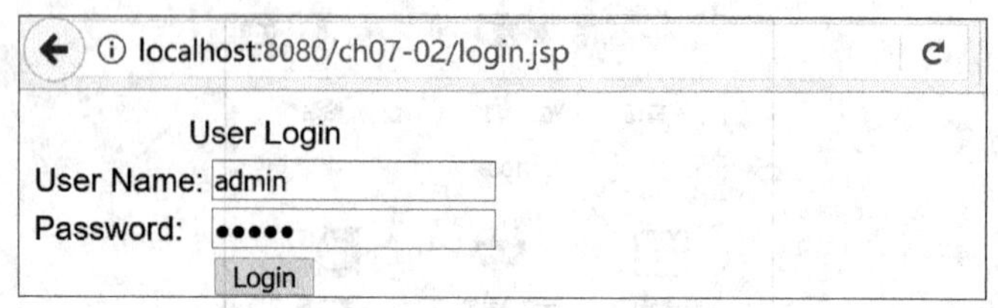

图 7-22　使用 Locale 为“英语/美国[en-us]”后刷新登录界面的效果

4. <fmt:param>

<fmt:param>标签用于在消息中传递动态参数。<fmt:param>必须位于<fmt:message>标签内部。

语法格式 1：使用 value 属性设置动态参数。

```
<fmt:param value="messageParameter"/>
```

语法格式 2：使用 fmt 的 body 设置动态参数。

```
<fmt:param >
    body
</fmt:param>
```

<fmt:param>的属性说明见表 7-26。

表 7-26　<fmt:param>的属性说明

属性	类型	是否必需	说明	默认值
value	Object	否	指定动态参数	无

下面说明使用<fmt:param>为动态参数赋值。首先在 Resources_en_US.properties 中增加键值对：

```
hello1=Hello\!{0},now is {1,date}
```

在 Resources_zh_CN.properties 增加键值对：

```
hello1=你好\!{0}。现在是 {1,date}
```

其中，Hello\! 是静态文本，{0}是第一个动态参数，在应用运行时被替换并动态显示；now is 是静态文本，{1,date}是第二个动态参数，动态参数类型是 date 类型。键 hello1 的值 Hello\! {0},now is {1,date}相当于一个含有动态参数的消息模板，动态参数在程序运行时刻动态地被替换。

接着编写 fmt_param.jsp，其关键代码如下：

```
<%request.setAttribute("time",new new java.util.Date(); %>
<fmt:bundle basename="Resources">
    <fmt:message key="hello1" >
```

```
        <fmt:param value="黎明"/>
        <fmt:param value="${time}"/>
    </fmt:message>
</fmt:bundle>
```

使用<fmt:param>设置第一个动态参数的值是黎明，设置第二个动态参数的值是java.util.Date()对象。在浏览器中 fmt_param.jsp 显示效果如图 7-23 所示。

图 7-23 fmt_param.jsp 显示效果

7.3.3 数字、时间日期格式化

1. <fmt:formatNumber>

<fmt:formatNumber>用于按照特定区域格式化并显示数字。例如，有一个数字 123456.789，区域设置为中文/中国时显示为 123,456.789；区域设置是德语/德国时显示为 123.456,789。

语法格式 1：没有 body，value 指定的值是待格式化的数字。

```
<fmt:formatNumber value="numericValue"
   [type="{number|currency|percent}"][pattern="customPattern"]
   [currencyCode="currencyCode"][currencySymbol="currencySymbol"]
   [groupingUsed="{true|false}"]
   [maxIntegerDigits="maxIntegerDigits"]
   [minIntegerDigits ="minIntegerDigits"]
   [maxFractionDigists ="maxFractionDigists"]
   [minFractionDigists ="minFractionDigists"]
   [var ="varName"]
   [scope ="{page|request|session|application}"] />
```

语法格式 2：没有 value，有 body，body 是待格式化的数字。

```
<fmt:formatNumber
   [type="{number|currency|percent}"][pattern="customPattern"]
   [currencyCode="currencyCode"][currencySymbol="currencySymbol"]
   [groupingUsed="{true|false}"]
   [maxIntegerDigits="maxIntegerDigits"]
   [minIntegerDigits ="minIntegerDigits"]
   [maxFractionDigists ="maxFractionDigists"]
   [minFractionDigists ="minFractionDigists"]
   [var ="varName"]
   [scope ="{page|request|session|application}"] >
```

```
待格式化的数字
</fmt:formatNumber>
```

如果<fmt:formatNumber>的 body 指定了待格式化的数字，那么 value 属性就不是必需的；如果没有 body 那么 value 就是必需的。<fmt:formatNumber>的属性说明见表 7-27。

表 7-27 <fmt:formatNumber>的属性说明

属性	是否必需	属性说明	默认值
value	否	指定待格式化的数字	无
type	否	指定格式化类型	number
pattern	否	指定格式化模式	取决于默认区域
currencyCode	否	指定货币代码	取决于默认区域
currencySymbol	否	指定货币符号，例如 $、¥	无
groupUsed	否	指定是否分隔数字，取值为 true|false，例如 1,234,567 是分割数字	true
maxIntegerDigits	否	指定整数部分最多显示的位数	无
minIntegerDigits	否	指定整数部分最少显示的位数	无
maxFractionDigits	否	指定小数点后最多显示的位数	无
minFractionDigits	否	指定小数点后最少显示的位数	无
var	否	指定一个变量来保存已经格式化的数字	无
scope	否	指定 var 的 JSP 作用域	page

type 属性用于指定格式化类型。type 属性的取值说明见表 7-28。

表 7-28 type 属性的取值说明

type 取值	属性说明	范例(当 value 为 0.88 时)
number	标准数字	0.88
currency	当地货币码	¥0.88
percent	百分比	88%

fmt_formatNumber.jsp 使用</fmt:formatNumber>格式化数字、货币、百分比，其关键代码如下：

```
<fmt:formatNumber value="0.88"/><br/>
<fmt:formatNumber>
   0.88
</fmt:formatNumber><br/>
<fmt:formatNumber type="number">
   0.88
</fmt:formatNumber><br/>
```

```
<fmt:formatNumber type="currency">
    0.88
</fmt:formatNumber><br/>
<fmt:formatNumber type="percent">
    0.88
</fmt:formatNumber>
```

浏览器 Locale 为中文/中国[zh-cn]时 fmt_formatNumber.jsp 显示效果如图 7-24 所示。

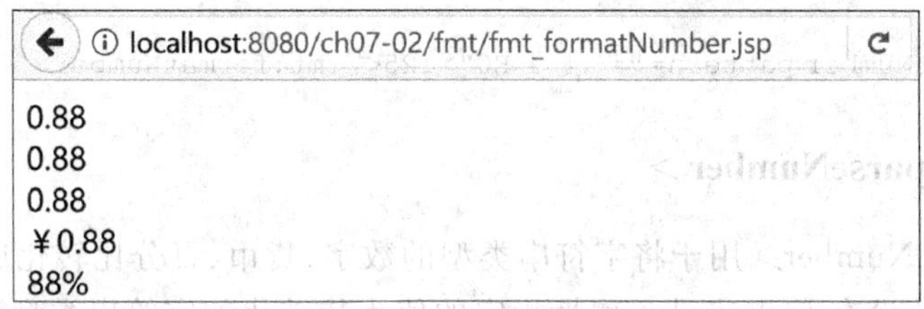

图 7-24　Locale 为中文/中国[zh-cn]时 fmt_formatNumber_type.jsp 显示效果

fmt_formatNumber2.jsp 使用 currencySymbol 属性指定货币符号，其关键代码如下：

```
<fmt:formatNumber type="currency" currencySymbol="$">
    0.88
</fmt:formatNumber>
```

仅当<fmt:formatNumber>的 type="currency"时，才能使用 currencyCode 和 currencySymbol 属性。currencyCode 属性用于在 type 属性为 currency 时指定货币代码，例如美元是 USD、人民币是 CNY；currencySymbol 属性用于指定货币符号，例如美元是 $，人民币是￥。

上面的代码使用 currencySymbol 属性指定货币符号是 $，这样不管客户的区域设置是什么，都会输出 $0.88。fmt_formatNumber2.jsp 的显示效果如图 7-25 所示。

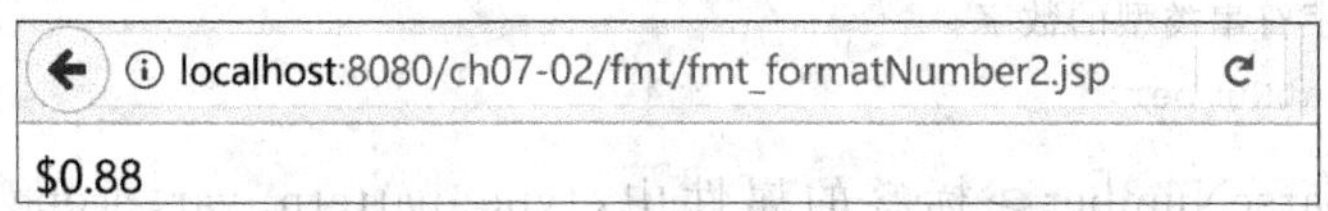

图 7-25　fmt_formatNumber2.jsp 的显示效果

groupingUsed 用于指定是否将数字进行分隔。当 groupingUsed 为 true 时，数字会进行分隔。例如，123456 分隔后是 123,456。groupingUsed 为 false 时，数字不进行分隔，例如 123456 不采用分隔，显示为 123456。

maxIntegerDigits 和 minIntegerDigits 用于指定数字的整数部分最多和最少显示的位数。当 maxIntegerDigits 小于待格式化数字的整数部分的位数时，会将整数部分多的位数部分删掉。例如数字 1234.56，当 maxIntegerDigits 是 3，那么结果是 234.56。当 minIntegerDigits 大于待格式化数字的整数部分时，会用 0 把不足的部分补齐。例如，待格式化的数字是 1234.56，minIntegerDigits 是 5，那么结果是 01234.56。

maxFractionDigits 和 minFractionDigits 用于指定待格式化数字的小数部分最多和最少显示的位数。当 maxFractionDigits 小于数字小数部分的位数时，会把多的部分删掉（删掉时采用四舍五入）。例如，当数字是 1.235 时，如果 maxFractionDigits 是 2，那么结果是 1.24。当 minFractionDigits 大于数字小数部分的位数时，会用 0 把不足的部分补齐，例如，当数字是 1.235 时，minFractionDigits 是 4，那么结果是 1.2350。

有时候显示数字会用到科学记数法。例如数字 125 用科学记数法表示 1.25E2，表示 1.25×10^2，可以用以下代码实现：

```
<fmt:formatNumber pattern="#.##E0">125</fmt:formatNumber>
```

2. <fmt:parseNumber>

<fmt:parseNumber>用于将字符串类型的数字、货币、百分比转化成数字。

语法格式 1：没有 body，value 属性指定的值为待转化的字符串类型数字。

```
<fmt:parseNumber value="numericValue" [type="{number|currency|percent}"]
   [pattern="customPattern"]
   [parseLocale="parseLocale"]
   [integerOnly="{true|false}"]
   [var="varName"]
   [scope="{page|request|session|application}"] />
```

语法格式 2：body 为待转化的字符串类型的数字。

```
<fmt:parseNumber value="numericValue" [type="{number|currency|percent}"]
   [pattern="customPattern"]
   [parseLocale="parseLocale"]
   [integerOnly="{true|false}"]
   [var="varName"]
   [scope="{page|request|session|application}"]>
   待转化的字符串类型的数字
</fmt:formatNumber>
```

在<fmt:parseNumber>标签的属性中，type、pattern、var、scope 属性的含义与<fmt:formatNumber>的对应属性相同，其他的属性说明见表 7-29。

表 7-29 <fmt:parseNumber>标签的其他属性说明

属性	是否必需	说明	默认值
value	否	指定待转化的字符串类型数字	无
parseLocale	否	指定格式化使用的区域设置	无
integerOnly	否	指定是否只显示整数部分	false

fmt_parseNumber.jsp 使用<fmt:parseNumber>的 value 属性指定待转化的字符串数字，其关键代码如下：

```
<fmt:parseNumber value="123456.789" />
```

显示结果为:123456.789。

fmt_parseNumber2.jsp 使用 body 指定待转化的字符串数字,其关键代码如下:

```
<fmt:parseNumber>123456.789</fmt:parseNumber>
```

显示结果为:123456.789。

type 属性用于指定将字符串解析成的类型,可以是 number(数字)或者 currency(货币)或者 percent(百分比),默认是数字。

fmt_parseNumber3.jsp 设置 type 属性指定为 percent 将字符串解析成货币,其关键代码如下:

```
<fmt:parseNumber type="percent">88%</fmt:parseNumber>
```

显示结果为:0.88。

integerOnly 用于指定转换成数字后是否保留小数部分。如果 integerOnly 为 true,则不会保留小数部分;如果 integerOnly 为 false,则会保留小数部分。parseLocale 属性用于指定一个临时的区域,根据该区域将字符串转化成数字。

fmt_parseNumber3.jsp 使用 integerOnly 和 parseLocale 属性,将字符串根据区域 en_US 转化成数字,转化后的数字不保留小数部分,其关键代码如下:

```
<fmt:parseNumber type="currency" parseLocale="en_US" integerOnly="true">
   $18.8
</fmt:parseNumber>
```

显示结果为:18。

3. <fmt:formatDate>

用于对 java.util.Date 类型的时间日期对象格式化并显示。

语法格式:

```
<fmt:formatDate value="datetime" [type="{time|date|both}"]
  pattern="customPattern"
  [dateStyle="{default|short|medium|long|full}"]
  [timeStyle="{default|short|medium|long|full}"]
  [timeZone="timeZone"]
  [var="varName"]
  [scope="{page|request|session|application}"] />
```

<fmt:formatDate>标签没有带 body 的语法格式,其属性说明见表 7-30。

表 7-30 <fmt:formatDate>标签的属性说明

属性	是否必需	属性说明	默认值
value	是	指定待格式化的 java.util.Date 类型的时间日期对象	无

续表

属性	是否必需	属性说明	默认值
type	否	指定格式化类型，取值为 date|time|both	date
dateStyle	否	指定日期部分的显示的样式，取值为 full|long|medium|short|default	default
timeStyle	否	指定时间部分的显示的样式，取值为 full|long|medium|short|default	default
pattern	否	指定格式化时间日期的模式	无
timeZone	否	指定格式化时采用的时区，取值为 EST|CST|MST|PST	默认时区
var	否	指定一个变量来保存格式化后的时间日期	无
scope	否	指定 var 的 JSP 作用域	page

fmt_formatDate.jsp 使用 value 属性格式化时间日期，其关键代码如下：

```
<jsp:useBean id="now" class="java.util.Date" />
<fmt:formatDate value="${now}"/><br/>
<fmt:formatDate value="${now}" type="time"/><br/>
<fmt:formatDate value="${now}" type="date"/><br/>
<fmt:formatDate value="${now}" type="both"/><br/>
```

<jsp:useBean>创建的时间日期对象 now 代表了程序运行的时间日期。

type 属性用于指定格式化类型，type="time"表示仅仅格式化时间，type="date"表示仅仅格式化日期，type="both"表示格式化日期和时间，默认是 date。

<fmt:formatDate value="${now}"/> 没有使用 type 属性，这时 type 属性取值为默认值 date，表示仅对日期格式化。后面 3 行分别使用 type 对时间格式化、对日期格式化、对时间和日期格式化。

如果浏览器区域设置是中文/中国[zh_CN]时，fmt_formatDate.jsp 运行结果可能为：

```
2017-11-18
11:24:23
2017-11-18
2017-11-18 11:24:23
```

fmt_formatDate2.jsp 使用 dateStyle 和 timeStyle 属性分别指定日期格式化样式和时间格式化样式，其关键代码如下：

```
<jsp:useBean id="now" class="java.util.Date" />
<fmt:formatDate value="${now}" type="date" dateStyle="default"/><br/>
<fmt:formatDate value="${now}" type="date" dateStyle="short"/><br/>
<fmt:formatDate value="${now}" type="date" dateStyle="medium"/><br/>
<fmt:formatDate value="${now}" type="date" dateStyle="long"/><br/>
```

```
<fmt:formatDate value="${now}" type="date" dateStyle="full"/><br/>
<fmt:formatDate value="${now}" type="time" timeStyle="default"/><br/>
<fmt:formatDate value="${now}" type="time" timeStyle="short"/><br/>
<fmt:formatDate value="${now}" type="time" timeStyle="medium"/><br/>
<fmt:formatDate value="${now}" type="time" timeStyle="long"/><br/>
<fmt:formatDate value="${now}" type="time" timeStyle="full"/><br/>
```

如果区域设置是中文(中国),fmt_formatDate2.jsp 的运行结果可能如表 7-31 所示。

表 7-31 区域设置是中文(中国)时 fmt_formatDate2.jsp 的运行结果

dateStyle 或 timeStyle 样式	日期结果	时间结果
default	2017-11-18	11:24:23
short	17-11-18	上午 11:24:23
medium	2017-11-18	11:24:23
long	2017 年 11 月 18 日	上午 11 时 24 分 23 秒
full	2017 年 11 月 18 日 星期六	上午 11 时 24 分 23 秒 CST

fmt_formatDate3.jsp.jsp 使用 pattern 属性指定时间日期格式化的模式,其关键代码如下:

```
<jsp:useBean id="now" class="java.util.Date" />
<fmt:formatDate value="${now}" type="both" timeStyle="full"
    pattern="yyyy.MM.dd G `at` HH:mm:ss z"/><br/>
<fmt:formatDate value="${now}" type="both" timeStyle="full"
    pattern="h:mm a"/><br/>
```

fmt_formatDate3.jsp 的运行结果如图 7-26 所示。

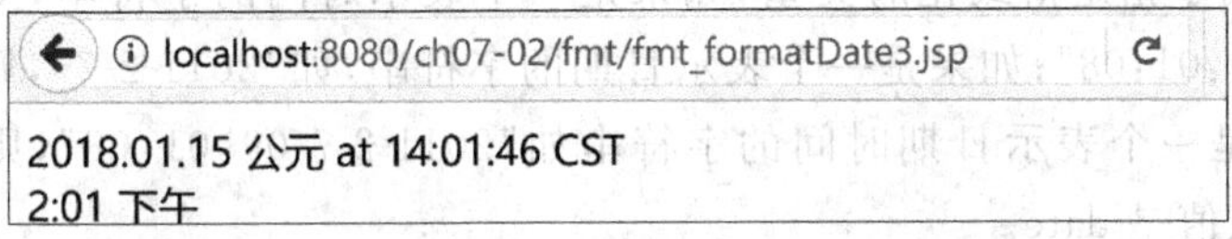

图 7-26 fmt_formatDate2.jsp 的运行结果

4. <fmt:parseDate>

<fmt:parseDate>标签用于将字符串类型的日期或时间转换成日期或时间类型。

语法格式 1:没有 body,value 指定待转换的字符串类型的日期或时间。

```
<fmt:parseDate value="datetime" [type="{time|date|both}"]
  [pattern="customPattern"][dateStyle="{default|short|medium|long|full}"]
  [timeStyle="{default|short|medium|long|full}"][timeZone="timeZone"]
  [var="varName"][scope="{page|request|session|application}"] />
```

语法格式 2:body 是待转换的字符串类型日期或时间。

```
<fmt:parseDate [type="{time|date|both}"]
```

```
    [pattern="customPattern"][dateStyle="{default|short|medium|long|full}"]
    [timeStyle="{default|short|medium|long|full}"][timeZone="timeZone"]
    [var="varName"][scope="{page|request|session|application}]>
    待转换的字符串类型的日期或时间
</fmt:parseDate>
```

<fmt:parseDate>标签的属性说明见表 7-32。

表 7-32 <fmt:parseDate>标签的属性说明

属性	是否必需	属性说明	默认值
value	是	指定待格式化的日期时间	无
type	否	指定格式化类型，取值为 date|time|both	date
dateStyle	否	指定日期部分的显示的样式	default
timeStyle	否	指定时间部分的显示的样式	default
pattern	否	指定格式化时间日期的模式	无
timeZone	否	指定格式化时间日期使用的时区	无
parseLocale	否	指定格式化时使用的区域设置	无
var	否	保存格式化后的日期时间	无
scope	否	var 的 JSP 作用域	page

fmt_parseDate.jsp 使用 value 属性指定待转换的字符串类型的日期或时间，其关键代码如下：

```
<fmt:parseDate value="2011-2-7"/>
```

显示结果为：Mon Feb 07 00:00:00 CST 2011。

type 属性用于指定格式化的类型，如果是一个表示时间的字符串，则必须指定 type="time"，如"09:01:08"；如果是一个表示日期的字符串，如"2011-2-7"，则必须指定 type="date"；如果是一个表示日期时间的字符串如"2011-2-7 09:01:08"，则必须指定 type="both"。默认值是 date。

fmt_parseDate2.jsp 使用 type 属性指定时间日期转换类型，其关键代码如下：

```
<fmt:parseDate value="09:01:08" type="time"/><br/>
<fmt:parseDate value="2011-2-7" type="date"/><br/>
<fmt:parseDate value="2011-2-7 09:01:08" type="both"/>
```

显示结果如下：

```
Thu Jan 01 09:01:08 CST 1970
Mon Feb 07 00:00:00 CST 2011
Mon Feb 07 09:01:08 CST 2011
```

fmt_parseDate3.jsp 使用 dateStyle 指定时间日期转换使用的日期样式。如果字符串为"2011 年 2 月 7 日 星期一"，那么字符串对应的 dateStyle 是 full，必须使用 dateStyle="full"，其关键代码如下：

```
<fmt:parseDate value="2011年2月7日星期一" type="date" dateStyle="full"/>
```

显示结果为：Mon Feb 07 00:00:00 CST 2011。

pattern 属性用于指定转换前字符串的样式。

fmt_parseDate4.jsp 使用 pattern 属性指定转换模式，其关键代码如下：

```
<fmt:parseDate value="20110207T162020" type="both"
 pattern="yyyyMMdd`T`HHmmss"/>
```

显示结果为：Mon Feb 07 16:20:20 CST 2011。

5. <fmt:setTimeZone>

<fmt:setTimeZone>标签用于设置时区，或将时区保存到某个 JSP 作用域中方便以后使用。

语法格式：

```
<fmt:setTimeZone value="timeZone"
   [var="varName"] [scope="{page|request|session|application}"] />
```

<fmt:setTimeZone>标签的属性说明见表 7-33。

表 7-33 <fmt:setTimeZone>标签的属性说明

属性	是否必需	属性说明	默认值
value	是	指定要设置的特定时区，取值可以是 EST、CST、MST、PST 等	无
var	否	指定一个变量来保存时区	无
scope	否	var 的 JSP 作用域	page

fmt_setTimeZone.jsp 设置时区为 EST 并保存在 session 作用域，其关键代码如下：

```
<c:set var="now" value="<%=new java.util.Date()%>" />
<p>当前时区时间：<fmt:formatDate value="${now}"
            type="both" timeStyle="long" dateStyle="long" /></p>
<p>修改为 GMT-8 时区:</p>
<fmt:setTimeZone value="GMT-8" />
<p>修改为 GMT-8 时区后的时间：<fmt:formatDate value="${now}"
            type="both" timeStyle="long" dateStyle="long" /></p>
```

fmt_setTimeZone.jsp 运行结果如下：

```
当前时区时间：2017年11月20日上午10时02分05秒
修改为 GMT-8 时区：
修改为 GMT-8 时区后的时间：2017年11月19日下午06时02分05秒
```

6. <fmt:timeZone>

<fmt:timeZone>标签用于设置一个临时的时区供其他标签使用。

语法格式：

```
<fmt:timeZone value="timeZone">
    body
</fmt:timeZone>
```

<fmt:timeZone>标签的属性说明见表 7-34。

表 7-34 <fmt:timeZone>标签的属性说明

属性	是否必需	属性说明	默认值
value	是	指定将要设置的临时时区。如果 value 为 null 或空时,默认是 GMT 时区	无

fmt_timeZone.jsp 使用<fmt:timeZone>设置一个临时时区,其关键代码如下:

```
<fmt:timeZone value="GMT">
<fmt:parseDate value="2011-02-01 22:00:00" type="both"
    var="datetime" scope="request"/>
</fmt:timeZone>
<fmt:formatDate value="${requestScope.datetime}" type="both" timeZone="GMT" />
```

<fmt:timeZone value="GMT">设置临时时区是 GMT。在<fmt:timeZone>的 body 中使用<fmt:parseDate>将"2011-02-01 22:00:00"转换成 GMT 时区的时间日期并保存在 request 作用域的 datetime 中;最后使用<fmt:formatDate>从 request 中取出 datetime 并转换成 GMT 时区的时间。

显示结果为:2011-2-1 22:00:00。

7.4 SQL 标签库

JSTL 的 SQL 标签库用于在 JSP 页面实现访问数据库的操作。在 JSP 页面使用 SQL 标签就是等于在表示层访问数据库,这不符合 MVC 模式。因此建议在大型项目中不用使用 SQL 标签。但是在一些小型项目中使用 JSP+SQL 标签可以快速开发应用。

SQL 标签库分为设置数据源和执行 SQL 语句。SQL 标签库分类见表 7-35。

表 7-35 SQL 标签库分类

标签库名称	功能分类	标签名称
SQL 标签库	设置数据源	setDataSource
	执行 SQL 语句	query dateParam paramtransaction update dateParam param

在 JSP 中使用 SQL 标签库时,<%@ taglib>指令通常如下所示:

```
<%@taglib uri="http://java.sun.com/jsp/jstl/sql" prefix="sql"%>
```

7.4.1 <sql:setDataSource>

<sql:setDataSource>标签用于引用一个数据源或设置一个数据源。

语法格式 1：引用一个已经存在的数据源。

```
<sql:setDataSource dataSource="dataSourceName"[var="varName"]
    [scope="page|request|session|application"] />
```

语法格式 2：设置一个数据源。

```
<sql:setDataSource url="databaseURL"user="dbUser"password="dbPassword"
    driver="driverName" [var="varName"] [scope="page|request|session|
application"] />
```

<sql:setDataSource>的属性说明见表 7-36。

表 7-36 <sql:setDataSource>标签的属性说明

属性	是否必需	属性说明	默认值
dataSource	否	指定数据源的名字	无
driver	否	指定 JDBC 驱动程序类名	无
url	否	指定数据库的 URL	无
user	否	指定数据库的用户名	无
password	否	指定数据库用户的密码	无
var	否	指定一个变量来保存数据源	无
scope	否	指定 var 的 JSP 作用域	page

例如，如下代码用于说明设置一个数据源，使用 var 属性指定数据源的名字是 mysqlDS 并保存在 session 作用域。

```
<sql:setDataSource
    url="jdbc:mysql://localhost:3306/testDB"
    driver="com.mysql.jdbc.Driver"
    user="root"password="root"scope="session"var="mysqlDS"/>
```

如下代码说明引用一个已经存在的数据源 mysqlDS，使用 var 为该数据源命名为 mysqlDS 并将其保存在 request 作用域。

```
<sql:setDataSource dataSource="mysqlDS"
    var="requestMysqlDS"scope="request"/>
```

7.4.2 <sql:query>

<sql:query>标签用于执行 select 语句。

语法格式：

```
<sql:query var="varName"[scope="{page|request|session|application}"]
    [dataSource="dataSource"]
    select 语句
</sql:query>
```

<sql:query>标签的属性说明见表 7-37。

表 7-37　<sql:query>标签的属性说明

属性	是否必需	属性说明	默认值
dataSource	否	指定使用的数据源	无
var	否	指定一个变量来保存查询结果	无
scope	否	指定 var 的 JSP 作用域	page

如果没有 dataSource 属性，会使用 page 范围默认的数据源。如果没有 dataSource 属性且 page 范围不存在默认数据源，那么会抛出异常。scope 属性用于指定保存查询结果的 var 变量的 JSP 作用域。

例如，如下代码用于查询数据库 testDB 中表 user 的所有记录。

```
<sql:setDataSource driver="com.mysql.jdbc.Driver"
  url="jdbc:mysql://localhost:3306/testDB"
  user="root" password="root" scope="session" var="mysqlDS"/>
<sql:query var="rs" dataSource="${sessionScope.mysqlDS}">
  select * from user
</sql:query>
<c:forEach items="${rs.rows}" var="row">
  ${row.name}
</c:forEach>
```

以上代码首先使用<sql:setDataSource>设置数据源；接着在<sql:query>中使用 dataSource 属性指定使用该数据源查询 user 表的所有记录，并将查询结果保存在 rs 中；最后使用<c:forEach>输出所有记录的 name 字段的值。rs.rows 表示使用字段名作索引的查询结果，可以通过 row.name 按照字段 name 取出该字段的值。

7.4.3　<sql:update>

<sql:update>标签用于执行 insert、update、delete 语句。

语法格式：

```
<sql:update [var="varName"] [scope="{page|request|session|application}"]
  [dataSource="dataSource"] >
  insert 或 update 或 delete 语句
</sql:update>
```

<sql:update>标签的属性说明见表 7-38。

表 7-38 <sql:update>标签的属性说明

属性	是否必需	属性说明	默认值
dataSource	否	指定数据源	无
var	是	指定一个变量来保存执行更新语句后受影响的记录行数	无
scope	否	指定 var 的 JSP 作用域	page

例如,如下代码使用<sql:update>进行更新操作。

```
<sql:setDataSource driver="com.mysql.jdbc.Driver"
   url="jdbc:mysql://localhost:3306/testDB"
   user="root" password="root" scope="session" var="mysqlDS"/>
<sql:update var="changedRows" dataSource="${sessionScope.mysqlDS}">
   update user set name=`Mike` where id=1
</sql:update>
changedRows=${pageScope.changedRows}
```

7.4.4 <sql:param>

<sql:query>或<sql:update>嵌套使用<sql:param>向 SQL 语句传递动态参数。

语法格式:

```
<sql:param>value</sql:param>
```

<sql:param>标签的属性见表 7-39。

表 7-39 <sql:param>标签的属性说明

属性	是否必需	属性说明	默认值
value	否	需要设置的参数值	无

例如,sql_param.jsp 使用<sql:param>传递参数,其关键代码如下:

```
<sql:setDataSource driver="com.mysql.jdbc.Driver"
   url="jdbc:mysql://localhost:3306/testDB"
   user="root" password="root" scope="session" var="mysqlDS"/>
<sql:update var="changedRows" dataSource="${sessionScope.mysqlDS}">
   delete from user where id=?
   <sql:param>1</sql:param>
</sql:update>
changedRows=${pageScope.changedRows}
```

SQL 语句 delete from user where id=? 的 id 是动态参数;<sql:param>1</sql:param>将参数 1 传递给 id。这将导致 id=1 的记录被删除。

7.4.5 <sql:dateParam>

<sql:dateParam>标签用于将 java. util. Date 转换为 java. sql. Date、java. sql. Time、java. sql. TimeStamp 类型。

语法格式：

```
<sql:dateParam type="{timestamp|time|date}"value="java.util.Date 的对象"/>
```

<sql:dateParam>标签的属性说明见表 7-40。

表 7-40 <sql:dateParam>标签的属性说明

属性	是否必需	说 明	默认值
value	否	指定待设置的时间日期参数(java. util. Date)	无
type	否	指定转换的目标时间日期类型， 取值为 time、date 或 timestamp	无

数据库中时间日期的数据类型通常有 date、time、timestamp，其对应的 Java 类型分别是 java. sql. Date、java. sql. Time、java. sql. TimeStamp。而 Java 应用中通常使用时间日期类型有 java. util. Date。这就需要将 java. util. Date 转换为 java. sql. Date、java. sql. Time、java. sql. TimeStamp。

例如，sql_dateParam. jsp 使用<sql:dateParam>将 java. util. Date 类型的对象转换为 java. sql. TimeStamp，其关键代码如下：

```
<sql:setDataSource driver="com.mysql.jdbc.Driver"
   url="jdbc:mysql://localhost:3306/testDB"
   user="root" password="root" scope="session" var="mysqlDS"/>
<jsp:useBean id="now" class="java.util.Date"/>
<c:out value="${now}"/>
<sql:update var="changedRows" dataSource="${sessionScope.mysqlDS}">
   update user set birthday =? where id =?
   <sql:dateParam type="timestamp" value="${now}"/>
   <sql:param>1</sql:param>
</sql:update>
changedRows=${pageScope.changedRows}
```

7.4.6 <sql:transaction>

<sql:transaction>标签用于设置使用数据库事务。

语法格式：

```
<sql:transaction [dataSource="dataSource"]
   [isolation="read_commited|read_uncommited|repeatable|serializable"] >
```

```
    <sql:query>或<sql:update>
</sql:transaction>
```

<sql:transaction>标签的属性说明见表 7-41。

表 7-41　<sql:transaction>标签的属性说明

属性	是否必需	说　　明	默认值
dataSource	否	数据源名称	无
isolation	否	事务隔离等级，取值为 READ_COMMITTED|READ_UNCOMMITTED| REPEATABLE_READ| SERIALIZABLE	数据库默认值

例如，sql_transaction.jsp 使用<sql:transaction>进行数据库事务操作，其关键代码如下：

```
<sql:setDataSource driver="com.mysql.jdbc.Driver"
  url="jdbc:mysql://localhost:3306/testDB"
  user="root" password="root" scope="session" var="mysqlDS"/>
<jsp:useBean id="now" class="java.util.Date"/>
<sql:transaction dataSource="${sessionScope.mysqlDS}">
  <sql:update>
     update user set birthday =?where id =?
     <sql:dateParam type="timestamp" value="${now}"/>
     <sql:param>1</sql:param>
  </sql:update>
  <sql:update>
     insert into user(name,password,age,birthday) values (?,?,?,?)
     <sql:param>Mike</sql:param>
     <sql:param>123</sql:param>
     <sql:param>23</sql:param>
     <sql:dateParam type="timestamp" value="${now}"/>
  </sql:update>
</sql:transaction>
```

以上代码在<sql:transaction>标签内部使用了两个<sql:update>标签分别进行更新和新增记录操作，这两个<sql:update>构成一个事务操作。

小　　结

JSP 中使用脚本段代码不利于角色分工，导致程序的可维护性和可扩展性较差。应当尽可能采用表达式语言和 JSTL 标签库来替代 JSP 中的脚本段。

JSTL 标签库实现了 JSP 常见 Web 操作。对于一个采用 MVC 模式的大中型应用，JSP 仅仅充当表示逻辑，开发中常常会用到 JSTL 的核心标签库中的<c:if>、

<c:forEach>、I18N 子标签库中的少数相关标签。对于仅仅使用 JSP 的小型应用，可能会用到 JSTL 标签库的大部分标签。

思考与习题

1. 什么是 JSTL?

2. JSTL 标签库是由哪些子标签库组成的?

3. 请写出<c:out>标签的语法格式，并举例说明使用<c:out>输出字符串 Hello、输出 JavaBean 对象 user 的 age 属性。

4. 请编写 JSP 页面实现用户注册，要求使用<c:set>将用户提交的注册数据设置到 JavaBean 对象 user 的属性中。

5. request 作用域中的 List 存放的是 Employe 类型的对象 employee。请编写一个 JSP 页面，使用<c:forEach>输出 employee 的 id 和 name 属性的值。

6. 请写出在<a href>标签中使用<c:url>实现跳转到百度首页。

7. 请编写一个 JSP 页面实现数字和时间日期格式化。

8. 请编写一个 JSP 页面，使用 SQL 标签库实现用户的增、删、改、查操作。

第8章

过滤器和监听器

Java Web 层组件除了 Servlet、JSP，还包括过滤器和监听器。

8.1 过 滤 器

过滤器本身并不生成请求和响应对象，它只是提供过滤作用。过滤器可以过滤的 Web 资源包括 Servlet、JSP 页面、HTML 页面、txt 文件、jpg 文件等。如果某个资源设置了过滤器，那么该资源文件被访问之前可以使用过滤器检查和修改请求对象；在该资源文件被访问之后检查和修改响应对象。可以为 Web 资源设置一个由多个过滤器组成的过滤器链并可以指定过滤器链中过滤器的顺序。

8.1.1 javax.servlet.Filter 接口

所有的过滤器必须实现 javax.servlet.Filter 接口。该接口包含过滤器必须实现的 3 个方法。

1. init(FilterConfig fConfig)

该方法是用于过滤器的初始化。Servlet 容器创建过滤器实例后会自动调用该方法。该方法的参数 FilterConfig 表示该过滤器的配置信息。

2. doFilter(ServletRequest request，ServletResponse response，FilterChain chain)

该方法完成实际的过滤操作。当一个 Web 资源文件被设置使用了过滤器时，Servlet 容器会首先调用该过滤器的 doFilter()方法。前两个参数分别是请求和响应对象。多个过滤器可以串联起来组成一个过滤器链对资源文件进行过滤，FilterChain chain 参数用于访问过滤器链中后续的过滤器。

3. destroy()

该方法用于释放过滤器占用的资源。Servlet 容器在销毁过滤器实例前会调用该方法。

8.1.2 配置过滤器

过滤器需要配置。配置过滤器有两种方式：使用 XML 配置和使用注解配置。下面说明创建过滤器类后分别使用 XML 和使用注解配置过滤器。

1. 创建过滤器类并在 web.xml 中配置

创建过滤器类 MyFilter，其关键代码如下：

```
public class MyFilter implements Filter {
    public void init(FilterConfig fConfig) throws ServletException {
    }
    public void doFilter(ServletRequest request, ServletResponse response,
        FilterChain chain) throws IOException, ServletException {
        System.out.println("调用 chain.doFilter(request, response)之前...");
        HttpServletRequest req = (HttpServletRequest) request;
        req.setAttribute("str", "a String");
        chain.doFilter(request, response);
        HttpServletResponse res = (HttpServletResponse) response;
        res.setHeader("refresh", "3");
        System.out.println("调用 chain.doFilter(request, response)之后...");
    }
    public void destroy() { }
}
```

MyFilter 过滤器类实现 javax.servlet.Filter 接口的 3 个抽象方法：init()、doFilter()、destroy()。在过滤器的核心方法 doFilter()中，req.setAttribute("str", "a String")用于在对请求中增加了一个 str 对象。chain.doFilter(request, response)将请求和响应转发给目标资源的下一个过滤器。如果目标资源没有后续过滤器，那么 chain.doFilter(request, response)会将请求和响应传递给目标资源。

执行了 chain.doFilter(request, response)后，对响应对象进行了修改，设置响应对象每隔 3s 进行刷新操作。这样被过滤的资源文件在被访问之后，客户端会执行刷新操作。

使用 XML 配置过滤器，需要在 web.xml 文件中使用<filter>和<filter-mapping>这两对标签。配置过滤器 MyFilter 的关键代码如下：

```
<web-app …>
   <filter>
      <filter-name>myFilter</filter-name>
```

```xml
        <filter-class>filter.MyFilter</filter-class>
    </filter>
    <filter-mapping>
        <filter-name>myFilterFilter</filter-name>
        <url-pattern>/index.jsp</url-pattern>
        <dispatcher>REQUEST</dispatcher>
    </filter-mapping>
    <filter-mapping>
        <filter-name>myFilterFilter</filter-name>
        <url-pattern>/another.jsp</url-pattern>
        <dispatcher>REQUEST</dispatcher>
    </filter-mapping>
</web-app>
```

<filter>用于声明一个过滤器。

<filter>标签有两个子标签<filter-name>和<filter-class>，分别指定过滤器名和过滤器类名（过滤器类的全限定名，即包名.类名）。

<filter-mapping>用于配置过滤器的过滤路径，即对哪些路径起到过滤作用。

<filter-mapping>的两个子标签<filter-name>和<url-pattern>，分别指定过滤器名和需要使用过滤器的资源路径。<filter-name>指定的名称必须与某个<filter>标签的子标签<filter-name>的名称一致。<url-pattern>指定需要使用过滤器的资源路径。

一个过滤器可以配置多个<filter-mapping>，表示该过滤器对多个映射路径进行过滤。以上使用<url-pattern>配置过滤器 myFilterFilter 对当前 Web 应用的根目录下的 index.jsp 和 another.jsp 均起过滤作用。

子标签<dispatcher>用于设置过滤器对应的请求方式。<dispatcher>设置可以是 REQUEST、FORWARD、INCLUDE、ERROR、ASYNC。如果没有配置<dispatcher>，默认是 REQUEST。

Servlet 2.3 规范中的过滤器只能由客户端发出请求来调用。如果使用 RequestDispatcher.forward()或 RequestDispatcher.include()方法来触发过滤器，过滤器不会执行。

Servlet 3.0 规范中的过滤器可以设置<dispatcher>为 REQUEST、FORWARD、INCLUDE、ERROR、ASYNC 来解决 Servlet 2.3 规范的上述不足。ASYNC 是指异步处理的请求可以触发过滤器。ERROR 是指由容器处理异常时转发的请求可以触发过滤器，例如，可以在 web.xml 中使用<error-page>指定出错页面 error.jsp 并为 error.jsp 配置过滤器 ErrorFilter；如果程序出现错误，在调用 error.jsp 之前会触发过滤器 ErrorFilter。

2. 创建过滤器类并在过滤器类中使用注解配置

创建过滤器类 myFilter，在类前直接使用@WebFilter 注解配置。

```java
@WebFilter(filterName="myFilter",value ={"/index.jsp","/another.jsp"},
```

```
        dispatcherTypes ={ DispatcherType.REQUEST })
public class MyFilter implements Filter {
    public void destroy() {}
    public void doFilter(ServletRequest request, ServletResponse response,
        FilterChain chain)throws IOException, ServletException {
        …
    }
    public void init(FilterConfig fConfig) throws ServletException {}
}
```

在过滤器类中使用注释配置属于 Servlet 3.0 规范新增内容。@WebFilter 的常用属性如下。

- filterName：String 类型，指定过滤器名，等价于<filter-name>。
- value：String[]类型，用于指定该过滤器需要过滤的多个 URL，该属性等价于 urlPatterns 属性。value 和 urlPatterns 不能同时使用。
- urlPatterns：String[]类型，用于指定一组过滤器的 URL 匹配模式。
- servletNames：String[]类型，用于指定过滤器将应用于哪些 Servlet。取值是 @WebServlet 中的 name 属性的值，或者是 web.xml 中<servlet-name>的值。
- dispatcherTypes：DispatcherType[]，指定过滤器的转发模式。取值为下列之一或多个：DispatcherType.REQUEST、DispatcherType.FORWARD、DispatcherType.INCLUDE、DispatcherType.ERROR、DispatcherType.ASYNC。默认是 DispatcherType.REQUEST。等价于 web.xml 中的<dispatcher>标记的作用。

在上述属性中，除了 value 或 urlPatterns 是必需的，其他都是可选的。

8.1.3 过滤器解决中文乱码

在 Java Web 应用中，中文乱码是常见问题之一。在中文乱码问题中，比较常见的是 Form 表单向服务器传送包含中文数据的时候会产生乱码。在过滤器出现之前，可以使用 request.setCharacterEncoding("utf-8")来解决，但是这会导致代码重复。现在可以使用过滤器 EncodingFilter 解决 Form 表单传送中文数据导致的乱码问题，其关键代码如下：

```
package filter;
//…导包略
public class EncodingFilter implements Filter {
    public void init(FilterConfig filterConfig) throws ServletException {}
    public void doFilter(ServletRequest request, ServletResponse response,
        FilterChain chain)throws IOException, ServletException {
        request.setCharacterEncoding("UTF-8");
        chain.doFilter(request, response);
    }
    public void destroy() {}
```

```
}
```

在过滤器的核心方法 doFilter()中，使用 request. setCharacterEncoding("UTF-8")设置请求编码为 UTF-8。chain. doFilter(request, response)将请求和响应转发给目标资源的下一个过滤器。如果目标资源没有后续过滤器，那么 chain. doFilter(request, response)会将请求和响应传递给目标资源。

在 web. xml 中配置过滤器 EncodingFilter，其关键代码如下：

```
</web-app…>
    <filter>
        <filter-name>encodingFilter</filter-name>
        <filter-class>filter.EncodingFilter</filter-class>
    </filter>
    <filter-mapping>
        <filter-name>encodingFilter</filter-name>
        <url-pattern>/*</url-pattern>
    </filter-mapping>
</web-app>
```

8.2 监 听 器

监听器是 Servlet 中一种特殊的类，能帮助开发者监听 Web 中的特定事件。例如，可以监听 Servlet 上下文、HTTP 会话(HttpSession)、请求(ServletRequest)的创建和销毁、属性的创建、销毁和修改等。

Servlet 2.4 规范定义的 Servlet 事件和监听器见表 8-1。

表 8-1 Servlet 2.4 规范定义的 Servlet 事件和监听器

监听器接口	事件类	监听器类别
ServletContextListener	ServletContextEvent	Servlet 上下文事件
ServletContextAttributeListener	ServletContextAttributeEvent	
HttpSessionListener	HttpSessionEvent	会话事件
HttpSessionActivationListener		
HttpSessionAttributeListener	HttpSessionBindingEvent	
HttpSessionBindingListener		
ServletRequestListener	ServletRequestEvent	请求事件
ServletRequestAttributeListener	ServletRequestAttributeEvent	

这些事件分为 3 类：Servlet 上下文事件、会话事件、请求事件。可以针对这些事件编写事件监听器，从而对事件做出相应处理。

8.2.1 监听器接口

1. javax. servlet. ServletContextListener

该接口包括如下两个抽象方法。

- contextInitialized(ServletContextEventsce)：用于监听 ServletContext 对象的创建。
- contextDestroyed(ServletContextEventsce)：用于监听 ServletContext 对象的销毁。

当创建 ServletContext 时，激发 contextInitialized(ServletContextEvent sce)方法；当销毁 ServletContext 时，激发 contextDestroyed(ServletContextEvent sce)。

例如，可以在 web. xml 中配置项目初始化信息后，在 contextInitialized()方法中获得初始化信息进行相关操作。在 web. xml 中配置项目初始化信息代码如下：

```
<web-app…>
   <context-param>
      <param-name>属性名</param-name>
      <param-value>属性值</param-value>
   </context-param>
</web-app>
```

在 contextInitialized 方法中获得初始化信息代码如下。

```
public class MyFirstListener implements ServletContextListener{
    public void contextInitialized(ServletContextEvent sce){
        //获取 web.xml 中配置的属性
        String value=sce.getServletContext().getInitParameter("属性名");
        System.out.println(value);
    }
    public void contextDestroyed(ServletContextEvent sce){
        //销毁 ServletContext 时的操作
    }
}
```

2. javax. servlet. ServletContextAttributeListener

该接口包括如下 3 个抽象方法。

attributeAdded(ServletContextAttributeEvent arg0)：用于监听对 ServletContext 属性的增加。

attributeReplaced(ServletContextAttributeEvent arg0)：用于监听对 ServletContext 属性的修改。

attributeRemoved(ServletContextAttributeEvent arg0)：用于监听对 ServletContext 属性

的删除。

3. javax. servlet. http. HttpSessionListener

该接口包括如下两个抽象方法。

- sessionCreated(HttpSessionEvent arg0)：用于监听 HttpSession 对象创建。
- sessionDestroyed(HttpSessionEvent arg0)：用于监听 HttpSession 对象销毁。

当创建一个 Session 时，激发 sessionCreated(SessionEvent se)方法；当销毁一个 Session 时，激发 sessionDestroyed(HttpSessionEvent se)方法。

4. javax. servlet. http. HttpSessionAttributeListener

该接口包括三个抽象方法。

- attributeAdded(HttpSessionBindingEvent arg0)：用于监听 HttpSession 对象中属性的增加。
- attributeReplaced(HttpSessionBindingEvent arg0)：用于监听 HttpSession 对象中属性的修改。
- attributeRemoved(HttpSessionBindingEvent arg0)：用于监听 HttpSession 对象中属性的删除。

当在 HttpSession 中增加一个属性时，会激发 attributeAdded(HttpSessionBindingEvent se)方法。

在 HttpSession 中属性被重新设置时，会激发 attributeReplaced(HttpSessionBindEvent se)方法。

在 HttpSession 中属性被删除时，会激发 attributeRemoved(HttpSessionBindingEvent arg0)。

5. javax. servlet. http. HttpSessionActiveListener

该接口包括两个抽象方法。

- sessionWillPassivate(HttpSessionEvent arg0)：用于监听 HttpSession 对象的激活。
- sessionDidActivate(HttpSessionEvent arg0)：用于监听 HttpSession 对象的钝化。

6. java. servlet. http. HttpSessionBindingListener

该接口包括两个抽象方法。

- valueBound(HttpSessionBindingEvent arg0)：用于监听 HttpSession 对象中对象绑定。
- valueUnbound(HttpSessionBindingEvent arg0)：用于监听 HttpSession 对象中删除。

7. javax.servlet.ServletRequestListener

该接口是在 Servlet 2.4 规范中新增的。该接口包括两个抽象方法。

- requestInitialized(ServletRequestEvent arg0)：用于监听请求的初始化。
- requestDestroyed(ServletRequestEvent arg0)：用于监听请求的销毁。

8. javax.servlet.ServletRequestAttributeListener

该接口包括三个抽象方法。

- attributeAdded(ServletRequestAttributeEvent arg0)：用于监听 ServletRequest 中属性的增加。
- attributeReplaced(ServletRequestAttributeEvent arg0)：用于监听 ServletRequest 中属性的修改。
- attributeRemoved(ServletRequestAttributeEvent arg0)：用于监听 ServletRequest 中属性的删除。

8.2.2 配置监听器

与过滤器配置类似，配置监听器也分两种方式：在使用 XML 配置和使用注解配置。

1. 创建监听器类并使用 XML 配置

监听器类 MyServletContextListener，其关键代码如下：

```
public class MyServletContextListener implements ServletContextListener {
    public void contextDestroyed(ServletContextEvent arg0) {      }
    public void contextInitialized(ServletContextEvent arg0) {
    }
}
```

MyServletContextListener 实现了 ServletContextListener 接口，因此必须实现其两个抽象方法 contextDestroyed(ServletContextEvent arg0)和 contextInitialized(ServletContextEvent arg0)。

在 web.xml 中，使用<listener>标签用于配置监听器类 MyServletContextListener，其关键代码如下：

```
<web-app …>
  <listener>
    <listener-class>
      listener.MyServletContextListener</listener-class>
    </listener>
</web-app>
```

<listener>的子标签<listener-class>指定监听器类的全限定名。

2. 创建监听器类并使用注解配置

```
@WebListener
public class MyServletContextListener implements ServletContextListener {
    public void contextDestroyed(ServletContextEvent arg0) {}
    public void contextInitialized(ServletContextEvent arg0) {}
}
```

在监听器类 MyServletContextListener 前面使用@WebListener 注释配置该类为一个监听器类。

8.2.3 监听器统计在线人数

Web 应用中经常需要统计应用的在线人数。可以使用 Servlet 监听器通过监听 HttpSession 的创建和销毁来统计在线人数。下面是一个简单的统计在线人数的 CounterListener 监听器,其关键代码如下:

```
package listener;
public class CounterListener implements HttpSessionListener {
    private static long onlineNumber =0;
    public static long getOnlineNumber() {
        return onlineNumber;
    }
    public void sessionCreated(HttpSessionEvent se) {
        onlineNumber++;
    }
    public void sessionDestroyed(HttpSessionEvent se) {
        onlineNumber--;
    }
}
```

CounterListener 监听器实现了 HttpSessionListener 接口,通过在 sessionCreated (HttpSessionEvent se)和 sessionDestroyed(HttpSessionEvent se)方法中,分别将表示在线人数的变量 onlineNumber 进行自增和自减操作,从而实现在线人数统计。

在 JSP 页面,可以使用如下语句显示在线人数:

```
当前共有<%=listener.CounterListener.getOnlineNumber()%>人在线<br>
```

小　　结

Servlet 过滤器是在 Servlet 2.3 规范中定义的。所有实现这一规范的 Servlet 容器都支持 Servlet 过滤器。过滤器本身并不生成请求和响应对象,它只是提供过滤作用。多个过滤器可以组成一个过滤器链且可以定顺序。配置过滤器可以使用 XML 配置或注解配置。过滤器主要应用包括认证、日志记录、图像转换、数据压缩、加密、字符串分解、XML 转换。Servlet 监听器对 Servlet 上下文事件、会话事件、请求事件进行监听。可以针对这些事件编写事件监听器，从而对事件做出相应处理。Servlet 监听器的主要应用包括统计在线用户、统计网站访问量、应用启动时进行初始化等。

思考与习题

1. 什么是 Servlet 过滤器？

2. javax.servlet.Filter 接口的 3 个方法的作用是什么？

3. 编写一个过滤器，实现过滤 Web 应用所有的资源。分别使用 XML 和注解配置该过滤器。

4. 监听器分为哪几类？分别监听何种事件？

5. 编写一个监听器，实现监听请求中属性的增加、删除和销毁。

6. 编写一个监听器，实现监听请求的初始化、销毁。

7. 编写一个监听器，实现在应用启动时读取 web.xml 中的数据库连接信息。

第9章

自定义标签

某些情况下 JSTL 无法满足要求,可以使用 JSP 自定义标签机制来创建定制标签库。

9.1 自定义标签简介

自定义标签是用户自定义的 JSP 元素。一个 JSP 自定义标签由标签处理器和标签描述两部分组成。标签处理器是实现 JspTag 接口的 Java 类。标签描述需要在标签库描述文件中描述标签和标签处理器。当 JSP 转换为 Servlet 并执行的时候,Web 容器会查找标签描述并将标签转换为对标签处理器实例的调用。

由于 JSP 2.0 新定义的标签与 JSP 1.2 的标签在实现原理上的不同,本书将 JSP 1.2 的标签称为传统标签,JSP 2.0 新定义的标签称为简单标签。JSP 2.0 既包含传统标签,也包含简单标签。JSP 2.0 自定义标签 API 的继承层次结构如图 9-1 所示。

JSP 自定义标签处理器需要实现 javax.servlet.jsp.tagext.JspTag 接口。该接口是一个标记接口,它有如下两个直接子接口。

Tag 接口:它是传统标签必须实现的接口。

SimpleTag 接口:这是 JSP 2.0 新增的接口,是简单标签需要实现的接口。

根据标签是否带属性和是否带标签体,标签可以分为如表 9-1 所示的类型。

表 9-1 根据标签是否带属性和是否带标签体的标签分类

标签分类名称	范　例
不带属性不带标签体的标签	<mytaglibs:certainTag/>
不带标签体但带有属性的标签	<mytaglibs:certainTag userName="Mike"/>
带标签体和属性的标签	<mytaglibs:certainTag userName="Mike"> 标签体内容 </mytaglibs:certainTag>
嵌套标签	<mytaglibs:certainTag> <mytaglibs:nestTag /> </mytaglibs:certainTag>

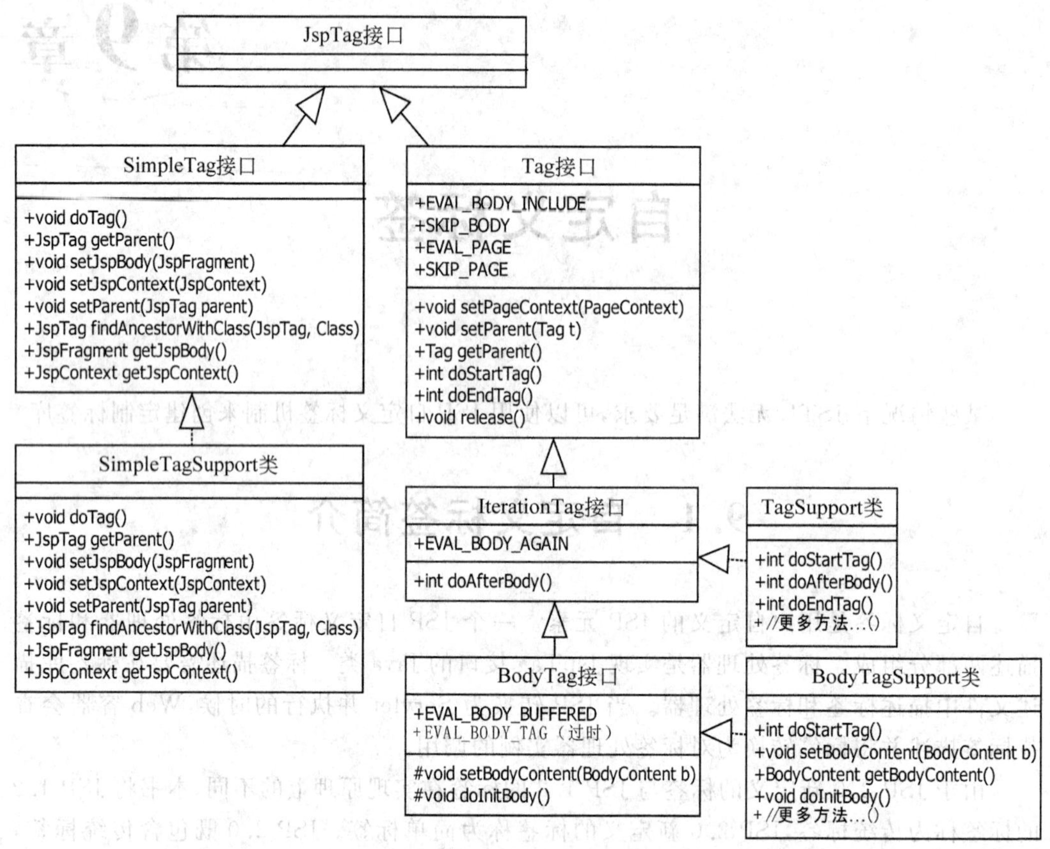

图 9-1　JSP 2.0 自定义标签 API 的继承层次结构

编写 JSP 自定义标签的基本步骤包括编写标签处理器、在标签库描述文件中描述标签。

9.2　传统标签

传统标签是指通过实现 javax.servlet.jsp.tagext.Tag 接口创建的标签。

传统标签的创建有两种方式：实现 Tag 相关接口；扩展 TagSupport 类或 BodyContent 类并覆盖相关方法。实际开发中通常采用第二种方式。

9.2.1　传统标签 API

JSP 传统标签 API 的继承层次结构如图 9-2 所示。

Tag 接口是传统标签必须实现的接口。它有一个直接子接口 IterationTag。IterationTag 接口用于开发实现迭代功能的标签；该接口有一个简单的实现类 TagSupport。实际开发中常常继承 TagSupport 类来创建迭代标签。

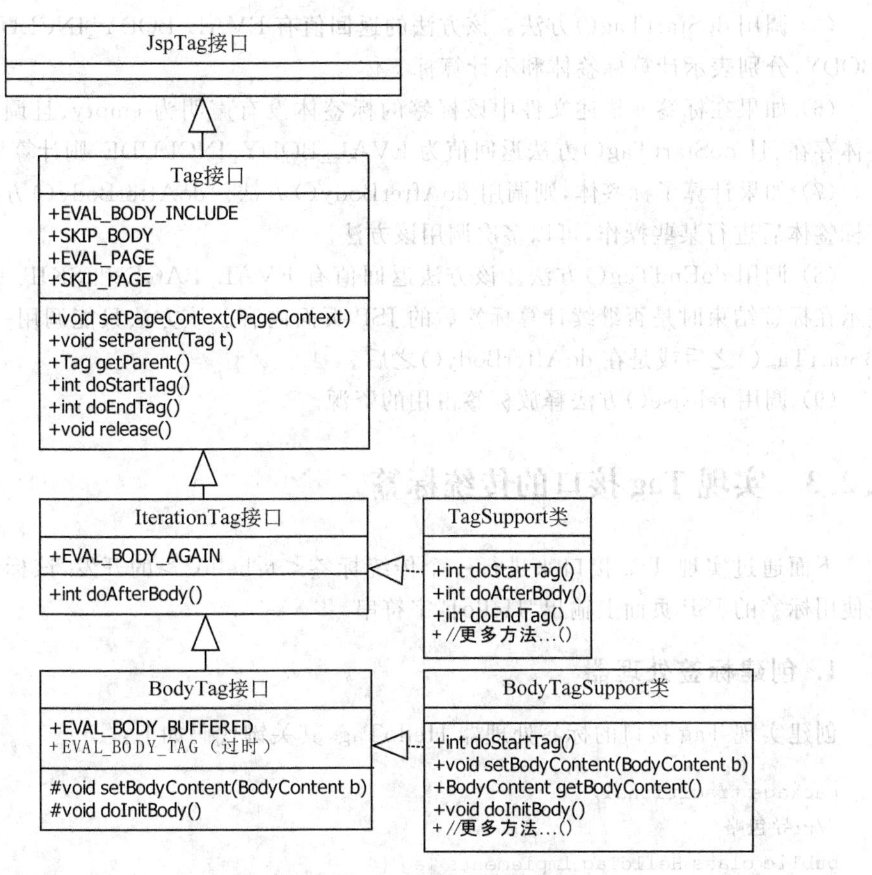

图 9-2　传统标签 API 继承层次结构图

IterationTag 接口的子接口 BodyTag 允许开发带标签体的标签。BodyTag 接口有一个实现类 BodyTagSupport。实际开发中通常直接继承 BodyTagSupport 来创建带标签体的标签。

9.2.2　传统标签生命周期

Tag 接口中定义了的方法与标签生命周期有关。传统标签的生命周期如下：

(1) 容器创建一个标签实例。容器通过调用标签处理器的无参数的构造方法完成标签实例创建。

(2) 容器调用 setPageContext(PageContext)方法。这样就为标签处理器提供 PageContext 的一个引用。

(3) 如果标签有父标签，则调用 setParent(Tag)方法设置标签的父标签；如果没有父标签则设置为 null。

(4) 如果标签有属性，则调用标签属性的 setter()方法。标签属性是在标签库描述文件中定义的、且在标签处理器中存在该属性及其 setter()方法；如果标签没有属性，就不用调用此方法。

(5) 调用 doStartTag()方法。该方法的返回值有 EVAL_BODY_INCLUDE 和 SKIP_BODY,分别表示计算标签体和不计算标签体。

(6) 如果在标签库描述文件中该标签的标签体没有声明为 empty、且调用标签时标签体存在、且 doStartTag()方法返回值为 EVAL_BODY_INCLUDE,则计算标签体。

(7) 如果计算了标签体,则调用 doAfterBody()方法。doAfterBody()方法允许在计算标签体后进行某些操作,可以多次调用该方法。

(8) 调用 doEndTag()方法。该方法返回值有 EVAL_PAGE 和 SKIP_PAGE,分别表示在标签结束时是否继续计算标签后的 JSP 页面内容。该方法只能调用一次,或是在 doStartTag()之后或是在 doAfterBody()之后。

(9) 调用 release()方法释放标签占用的资源。

9.2.3 实现 Tag 接口的传统标签

下面通过实现 Tag 接口来进行一个传统标签<t:hello/>的开发,该标签的功能是在使用标签的 JSP 页面上输出"Hello!"字符串。

1. 创建标签处理器

创建实现 Tag 接口的标签处理器 HelloTag,其关键代码如下:

```
package traditional;
//…导包略
public class HelloTag implements Tag {
    private PageContext pageContext;
    private Tag parent;
    @Override
    public int doEndTag() throws JspException {
        try {
            this.pageContext.getOut().println("Hello!");
        } catch (IOException e) {
            e.printStackTrace();
        }
        return EVAL_PAGE;
    }
    @Override
    public int doStartTag() throws JspException {
        return SKIP_BODY;
    }
    @Override
    public Tag getParent() {
        return parent;
    }
    @Override
```

```
    public void release() {}
    @Override
    public void setPageContext(PageContext pageContext) {
        this.pageContext =pageContext;
    }
    @Override
    public void setParent(Tag parent) {
        this.parent =parent;
    }
}
```

2. 编写标签库描述文件

在 Web 项目的 WEB-INF\tlds 下新建标签库描述文件 traditional.tld 文件，代码如下：

```
<?xml version="1.0" encoding="UTF-8" ?>
<taglib xmlns="http://java.sun.com/xml/ns/j2ee"
  xmlns:xsi="http://www.w3.org/2001/XMLSchema-instance"
  xsi:schemaLocation=http://java.sun.com/xml/ns/j2ee
  http://java.sun.com/xml/ns/j2ee/web-jsptaglibrary_2_0.xsdversion="2.0">
  <description>mytraditionaltag</description>
  <tlib-version>1.0</tlib-version>
  <short-name>t</short-name>
  <uri>/mytraditionaltag</uri>
  <tag>
    <description>实现 Tag 接口，输出 Hello!</description>
    <name>hello</name>
    <tag-class>traditional.HelloTag</tag-class>
    <body-content>empty</body-content>
  </tag>
</taglib>
```

标签库描述文件用于配置标签库和标签。该文件中的主要标签说明如下。

- <taglib>：标签库描述文件的根标签。
- <description>：可选，指定标签库描述信息。
- <tlib-version>：指定标签库的版本。
- <short-name>：指定标签库的简写名称。通常 JSP 页面中<taglib>的 prefix 前缀与该简写名称是一致的。
- <uri>：指定标签库的网络位置。
- <tag>标签：描述某个自定义标签。
- <tag>标签的<description>：可选。指定某个标签的描述信息。
- <tag>标签的<name>：指定标签名。

- <tag>标签的<tag-class>：指定标签类。
- <tag>标签的<body-content>：取值为 empty、JSP、scriptless 和 tagdependent 之一。指定标签是否有标签体。
- empty：表示没有标签体。
- JSP：表示标签体可以包含合法的 JSP 元素。
- scriptless：表示标签体可以包含 EL 表达式和 JSP 动作元素，但不能包含 JSP 的脚本元素(包括脚本段代码、表达式和声明)。
- tagdependent：表示在标签体中所写的任何代码都会原封不动地传给标签处理器。

JSP 2.0 不需要在 web.xml 中配置标签库。JSP 1.2 需要在 web.xml 中配置标签库，在此不进行说明。

在 JSP 页面 HelloTag.jsp 使用自定义标签，其关键代码如下：

```
...
<%@taglib uri="/WEB-INF/tlds/traditional.tld" prefix="t" %>
<html><head></head>
<body>
   <t:hello/>
</body>
</html>
```

9.2.4 继承 TagSupport 类的传统标签

下面通过继承 TagSupport 类来完成一个标签<t:helloClass/>的开发。该标签的功能是在 JSP 页面输出"HelloTagSupport!"字符串。

1. 创建标签处理器

创建继承 TagSupport 类的标签处理器 HelloTagClass，其关键代码如下：

```
package traditional;
//…导包略
public class HelloTagClass extends TagSupport {
    @Override
    public int doStartTag() throws JspException {
        try {
            pageContext.getOut().write("Hello TagSupport!");
        } catch (IOException e) {
            throw new JspException("IOException" +e.toString());
        }
        return SKIP_BODY;
    }
```

```
}
```

继承 TagSupport 类创建标签处理器中仅仅覆盖了 doStartTag()方法。与实现 Tag 接口比较，继承 TagSupport 创建标签方式较为简单。

2. 编写标签库描述文件

在 Web 项目的 WEB-INF\tlds 的 traditional. tld 文件中，新增代码如下：

```
<tag>
  <description>hello tag by implements tagsupport</description>
  <name>helloClass</name>
  <tag-class>traditional.HelloTagClass</tag-class>
  <body-content>empty</body-content>
</tag>
```

在 JSP 页面 HelloTagClass. jsp 使用自定义标签，其关键代码如下：

```
<%@taglib uri="/WEB-INF/tlds/traditional.tld" prefix="t" %>
<html><head></head>
<body>
   <t:helloClass/>
</body>
</html>
```

9.2.5 带属性和标签体的传统标签

有时候需要创建带标签体和属性的标签。创建带标签体的标签有两种方式：实现 javax. servlet. jsp. tagext. BodyTag 接口；继承 javax. servlet. jsp. tagext. TagSupport 类。第二种方式较第一种方式简单，实际开发中通常直接继承 TagSupport 类来创建迭代功能的标签。

下面介绍使用第二种方式实现迭代功能的标签＜t:iteration items＝" $ {items }" var＝"name"＞，该标签的属性 items 和 name 分别是 String[]和 String 类型。

1. 创建标签处理器

继承 TagSupport 实现迭代功能的 IterationTag 标签类，其关键代码如下：

```
package traditional;
//…导包略
public class IterationTag extends TagSupport {
    private String var;
    private String[] items;
    private int i =1;
    public void setVar(String var) {
```

```java
        this.var =var;
    }
    public void setItems(String[] items) {
        this.items =items;
    }
    @Override
    public int doStartTag() throws JspException {
        if (null !=items && items.length >0) {
            pageContext.setAttribute("name", items[0]);
            return EVAL_BODY_INCLUDE;
        } else {
            return SKIP_BODY;
        }
    }
    @Override
    public int doAfterBody() throws JspException {
        if (i <items.length) {
            pageContext.setAttribute("name", items[i]);
            i++;
            return EVAL_BODY_AGAIN;
        } else {
            return SKIP_BODY;
        }
    }
    @Override
    public int doEndTag() throws JspException {
        return EVAL_PAGE;
    }
}
```

IterationTag 类中的实例变量 String[] items、String var 表示标签属性，在标签处理器中属性都有对应的 setter 方法。这两个属性分别表示待迭代的 String 数组和每次迭代时数组中的一个元素。标签属性需要在标签库描述文件中描述。int i 表示迭代次数。

该标签处理器中覆盖了父类的 3 个方法：doStartTag()、doAfterBody()、doEndTag()。实现迭代的操作需要在 doStartTag()中返回 EVAL_BODY_INCLUDE，在 doAfterBody()中返回 EVAL_BODY_AGAIN。由于此标签处理器需要迭代一个 String 数组，并在标签体中访问数组元素，故需要在 doStartTag()方法中首先判断数组存在且数组元素个数大于 0，满足条件会将第一个数组元素通过 pageContext. setAttribute("name", items[i])放入 page 范围并返回表示计算标签体的常量 EVAL_BODY_INCLUDE；不满足条件返回 SKIP_BODY 表示不计算标签体。在 doAfterBody()中判断迭代次数 i，如果小于数组长度会重复进行计算标签体工作；否则不计算标签体。doEndTag()返回 EVAL_PAGE 表示计算标签后的剩余 JSP 页面。

2. 编写标签库描述文件

在 Web 项目的 WEB-INF\tlds 的 traditional. tld 文件中，新增代码如下：

```
<tag>
  <description>IterationTag</description>
  <name>iteration</name>
  <tag-class>traditional.IterationTag</tag-class>
  <body-content>scriptless</body-content>
  <attribute>
    <name>var</name>
    <required>true</required>
  </attribute>
  <attribute>
    <name>items</name>
    <required>true</required>
    <rtexprvalue>true</rtexprvalue>
  </attribute>
</tag>
```

<attribute>标签指定当前标签的两个属性 var 和 items。items 属性中<rtexprvalue>true</rtexprvalue>，表示可以使用 JSP 脚本元素和 EL 表达式来为属性动态设置值。

使用标签的 JSP 页面 iterationTag. jsp 的关键代码如下：

```
<body>
<%
    String[] items ={"1", "2", "3"};
    pageContext.setAttribute("items", items);
%>
<t:iteration items="${items }" var="name">
  ${name }
</t:iteration>
</body>
```

运行该 JSP 页面，会在页面上输出字符 1 2 3。

9.2.6 修改内容的传统标签

大部分情况下，Tag 和 IterationTag 接口的生命周期方法能满足要求。使用 doStartTag()、doAfterBody()、doEndTag()几乎可以完成任何功能的标签编写工作。但是，某些情况下需要访问标签体的内容。这时需要继承 BodyTagSupport 类。通过继承 BodyTagSupport 类可以从 BodyTag 接口得到其他两个生命周期方法：setBodyContent()和 doInitBody()。通过使用这些方法处理标签体内容。下面介绍通过继承 BodyTagSupport 类来完成修改标签体内容的如下标签：

```
<t:ModifyBodyContent>
  in ModifyBodyContent.jsp。
</t:ModifyBodyContent>
```

1. 创建标签处理器

通过继承 BodyTagSupport 的 ModifyBodyContent 标签处理器关键代码如下：

```
package traditional;
//…导包略
public class ModifyBodyContent extends BodyTagSupport {
    @Override
    public int doStartTag() throws JspException {
        return EVAL_BODY_BUFFERED;
    }
    @Override
    public void setBodyContent(BodyContent b) {
        super.setBodyContent(b);
    }
    @Override
    public int doEndTag() throws JspException {
        String content =bodyContent.getString();
        //System.out.println(content);
        String newStr ="新的字符串";
        JspWriter jspWriter =bodyContent.getEnclosingWriter();
        try {
            jspWriter.write(newStr);
        } catch (IOException e) {
            e.printStackTrace();
        }
        return EVAL_PAGE;
    }
}
```

ModifyBodyContentt 的 doStartTag()方法返回值为 EVAL_BODY_BUFFERED，这样导致容器接着会调用 setBodyContent(BodyContent b)。在 setBodyContent(BodyContent b)方法中 super.setBodyContent(b)设置实例变量 BodyContent bodyContent 为方法传入参数值，而 BodyContent bodyContent 是在 BodyTagSupport 中定义的。在 doEndTag()方法中获取 bodyContent 并得到其 String 内容，构建一个新的字符串 newStr，通过 bodyContent 获得 jspWriter、调用 jspWriter.write(newStr)方法将表示新字符串的 newStr 写到输出流。

2. 编写标签库描述文件

在 Web 项目的 WEB-INF\tlds 的 traditional.tld 文件中，新增代码如下：

```
<tag>
  <description>ModifyBodyContent</description>
  <name>ModifyBodyContent</name>
  <tag-class>traditional.ModifyBodyContent</tag-class>
  <body-content>scriptless</body-content>
</tag>
```

JSP 页面 ModifyBodyContent.jsp 使用标签，其关键代码如下：

```
<body>
<t:ModifyBodyContent>
   in ModifyBodyContent.jsp。
</t:ModifyBodyContent>
</body>
```

运行 ModifyBodyContent.jsp，页面会显示"新的字符串"。

9.3 简单标签

简单标签的创建有两种方式：实现 SimpleTag 接口；扩展 SimpleTag 接口的实现类 SimpleTagSupport 并覆盖其 doTag()方法。实际开发中通常采用第二种方式。

9.3.1 简单标签 API

简单标签 API 的继承层次结构如图 9-3 所示。最上层的 JspTag 接口仅仅是一个标识接口。SimpleTag 接口中定义的这些方法都是简单标签的生命周期方法。SimpleTagSupport 类实现了 SimpleTag 接口的方法。其中 doTag()方法不做任何操作，因此需要在 SimpleTagSupport 的子类中覆盖该方法。最下方的 3 个方法是该类新增的，其中最实用的是 getJspBody()方法。

9.3.2 简单标签生命周期

JSP 页面调用一个简单标签时，Web 容器创建一个新的标签处理器实例，并调用该处理器上的两个或多个方法。当 doTag()方法完成，容器会销毁处理器实例。简单标签的生命周期如下：

(1) 容器创建一个简单标签实例。容器创建标签处理器实例是通过调用其无参数的构造方法完成的。

(2) 调用 setJspContext(JspContext)方法，为处理器提供 PageContext 引用。PageContext 是 JspContext 的子类。

(3) 如果标签有父标签，则调用 setParent(JspTag)方法。

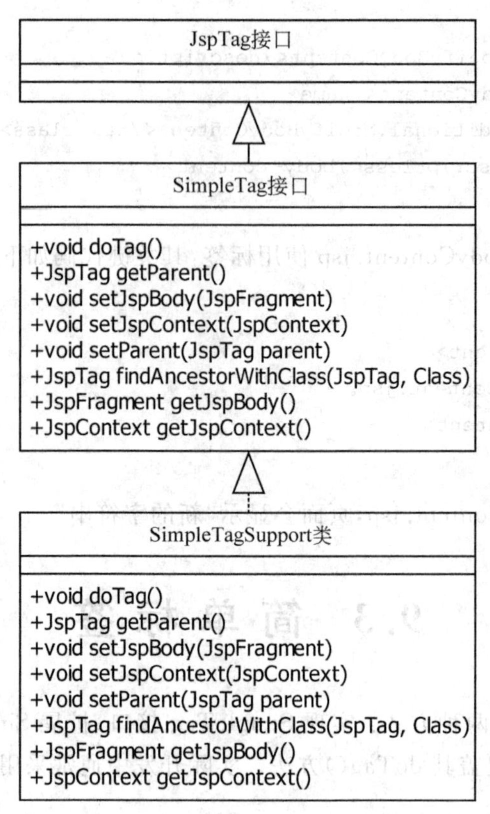

图 9-3　简单标签 API 继承层次结构

(4) 如果标签带有属性,则调用属性的 setter 方法。在带有属性的标签处理器中,需要手工增加该属性及其 setter 方法。

(5) 如果标签的<body-content>未声明为 empty 且标签有标签体,则调用 setJspBody(JspFragment)方法。

(6) 调用 doTag()方法。

9.3.3　继承 SimpleTagSupport 的简单标签

下面通过继承 SimpleTagSupport 类来创建一个简单标签<s:hello/>,该标签的作用是在 JSP 页面输出字符串"Hello!"。

1. 创建标签处理器

继承 SimpleTagSupport 类的 HelloSimpleTag 标签处理器的关键代码如下:

```
package simple;
//…导包略
public class HelloSimpleTag extends SimpleTagSupport{
    @Override
```

```
    public void doTag() throws JspException, IOException {
        getJspContext().getOut().print("Hello!");
    }
}
```

HelloSimpleTag 覆盖了父类 SimpleTagSupport 类的 doTag()方法。在 doTag()方法中，getJspContext()方法获得了一个 JSPContext 对象；JSPContext. getOut()方法用于获得一个 JspWriter 对象；最后调用 JspWriter. print()方法输出字符串"Hello!"。

2. 编写标签库描述文件

在标签库描述文件\WEB-INF\tlds2\simpletag. tld 中描述简单标签的代码如下：

```
<?xml version="1.0" encoding="UTF-8"?>
<taglib xmlns="http://java.sun.com/xml/ns/j2ee"
  xmlns:xsi="http://www.w3.org/2001/XMLSchema-instance"
  xsi:schemaLocation=http://java.sun.com/xml/ns/j2ee
  http://java.sun.com/xml/ns/j2ee/web-jsptaglibrary_2_0.xsdversion="2.0">
  <description>MySimpleTag-JSP2.0</description>
  <tlib-version>1.0</tlib-version>
  <short-name>s</short-name>
  <uri>/mysimpletag</uri>
  <tag>
      <name>hello</name>
      <tag-class>simple.HelloSimpleTag</tag-class>
      <body-content>empty</body-content>
  </tag>
</taglib>
```

JSP 页面 HelloSimpleTag. jsp 使用标签，其关键代码如下：

```
<%@taglib uri="/mysimpletag" prefix="s"%>
<html>
<head><title>HelloSimpleTag</title></head>
<body>
  <s:hello/>
</body>
</html>
```

9.3.4 输出标签体内容的简单标签

下面通过继承 SimpleTagSupport 类来创建一个简单标签：

```
<s:helloWithBody>HelloWithBody!</s:helloWithBody>
```

该标签的作用是在 JSP 页面输出标签体内容"HelloWithBody!"。

1. 创建标签处理器

继承 SimpleTagSupport 类的 HelloTagWithBody 标签类，其关键代码如下：

```
package simple;
//…导包略
public class HelloTagWithBody extends SimpleTagSupport{
    @Override
    public void doTag() throws JspException, IOException {
        getJspBody().invoke(null);
    }
}
```

HelloTagWithBody 标签处理器覆盖了父类的 doTag()方法。getJspBody()方法返回一个 JspFragment 对象，该对象的 invoke(Writer out)需要一个 Writer 类型的参数。invoke(null)表示将标签体内容输出到响应；如果 invoke()方法的参数是一个 Writer 类型表示将标签体内容作为参数传给 Writer。

2. 编写标签库描述文件

在标签库描述文件\WEB-INF\tlds2\simpletag. tld 中配置简单标签，其关键代码如下：

```
<tag>
  <name>helloWithBody</name>
  <tag-class>simple.HelloTagWithBody</tag-class>
  <body-content>scriptless</body-content>
</tag>
```

<body-content>配置标签体为 scriptless，表示标签可能有标签体，但是标签体中不能有脚本(脚本段代码、表达式、声明)

在 JSP 页面 HelloTagWithBody. jsp 使用标签，其关键代码如下：

```
<body>
  <s:helloWithBody>HelloTagWithBody!</s:helloWithBody>
</body>
```

9.3.5 带属性的简单标签

下面通过继承 SimpleTagSupport 类来创建一个简单标签：

```
<s:withAttribute name="SomeValue"/>
```

该标签的作用是在 JSP 页面输出标签属性值。例如 name="Mike"，会在 JSP 页面输出 Mike。

1. 创建标签处理器

继承 SimpleTagSupport 类的 WithAttribute 标签处理器,其关键代码如下:

```
package simple;
//…导包略
public class WithAttribute extends SimpleTagSupport {
    private String name;
    public void setName(String name) {
        this.name =name;
    }
    @Override
    public void doTag() throws JspException, IOException {
        getJspContext().getOut().print(this.name);
    }
}
```

WithAttribute 中定义了表示标签属性的 name 字段及其 setter()方法。在覆盖了父类的 doTag()方法中,getJspContext(). getOut(). print(this. name)表示在 JSP 页面输出 name 的值。

2. 编写标签库描述文件

在标签库描述文件\WEB-INF\tlds2\simpletag. tld 中配置简单标签的关键代码如下:

```
<tag>
  <name>withAttribute</name>
  <tag-class>simple.WithAttribute</tag-class>
  <body-content>empty</body-content>
  <attribute>
    <name>name</name>
    <required>true</required>
    <rtexprvalue>true</rtexprvalue>
  </attribute>
</tag>
```

<body-content>为 empty 表示此标签没有标签体。

<attribute>表示配置标签属性。其子标签<name>配置标签属性名,这里标签属性名是 name;<required>为 true 表示标签属性是必需的;<rtexprvalue>为 true 表示属性值可以使用 EL 表达式或 JSP 脚本元素动态获得。

创建在 JSP 页面 WithAttribute. jsp 使用标签,其关键代码如下:

```
<body>
<s:withAttribute name="Mike"/>
```

```
</body>
```

9.3.6 修改标签体内容的简单标签

下面通过继承 SimpleTagSupport 创建一个简单标签：

```
<s:modifybodycontent>bodyContent</s:modifybodycontent>
```

该标签的作用是在 JSP 页面输出标签处理器指定的内容 modified body Content，而不是原标签体内容 bodyContent。

1. 创建标签处理器

通过继承 SimpleTagSupport 类的 ModifyBodyContent 标签处理器的关键代码如下：

```
package simple;
//…导包略
public class ModifyBodyContent extends SimpleTagSupport {
    @Override
    public void doTag() throws JspException, IOException {
        StringWriter stringWriter =new StringWriter();
        JspFragment jspFragment =getJspBody();
        jspFragment.invoke(stringWriter);
        String content =stringWriter.toString();
        content ="modified body Content";
        PageContext pageContext = (PageContext) getJspContext();
        pageContext.getOut().write(content);
    }
}
```

在 ModifyBodyContent 的 doTag()方法中，首先创建了一个 StringWriter 对象；getJspBody()获得一个 JspFragment 对象 jspFragment。jspFragment. invoke(stringWriter)将标签体原有的内容指定到一个 stringWriter。最后，使用 PageContext 对象将标签体的新内容输出。

2. 编写标签库描述文件

在标签库描述文件\WEB-INF\tlds2\simpletag. tld 中配置简单标签，其关键代码如下：

```
<tag>
  <name>modifybodycontent</name>
  <tag-class>simple.ModifyBodyContent</tag-class>
  <body-content>scriptless</body-content>
</tag>
```

在 JSP 页面 WithAttribute.jsp 使用标签，其关键代码如下：

```
<s:modifybodycontent>
  bodyContent
</s:modifybodycontent>
```

小　　结

JSP 2.0 自定义标签包括 JSP 1.2 的传统标签和 JSP 2.0 的简单标签。与 JSP 1.2 的传统标签相比，JSP 2.0 的简单标签简化了标签编程。开发标签时尽可能采用 JSP 2.0 的标签。开发 JSP 标签可以通过相关实现接口或者通过继承相关实现类方式。通常建议采用继承实现类方式开发标签。

思考与习题

1. 什么是自定义标签库？

2. 根据标签是否带属性和标签体，举例说明标签的分类。

3. 什么是传统标签？开发一个传统标签可以实现哪些接口或继承哪些类？

4. 请写出传统标签的生命周期。

5. 通过实现 Tag 接口来开发一个传统标签<t:helloworld/>，实现 JSP 页面输出“你好，世界！”。

6. 通过继承 TagSupport 类来开发一个传统标签<t:helloworldClass/>，实现 JSP 页面输出“你好，世界！”。

7. 通过继承 TagSupport 类来创建迭代功能的标签<t:iteration items="${items}" var="var">，该标签的属性 items 和 var 分别是 List 和 Integer 类型。

8. 通过继承 BodyTagSupport 类来完成修改标签体内容的如下标签：

```
<t:ModifyBodyContent>
  in ModifyBodyContent.jsp。
</t:ModifyBodyContent>
```

将标签体内容修改为“New String！”。

9. 什么是简单标签？开发一个简单标签可以实现哪些接口或继承哪些类？

10. 请写出简单标签的生命周期。

11. 通过继承 SimpleTagSupport 类来创建一个简单标签<s:helloWorld/>，该标签的作用是在 JSP 页面输出“你好，世界！”。

12. 通过继承 SimpleTagSupport 类来创建一个简单标签：

```
<s:helloWithBody>SimpleTagSupport!</s:helloWithBody>
```

该标签的作用是在 JSP 页面输出标签体内容“SimpleTagSupport!”。

13. 通过继承 SimpleTagSupport 类来创建一个如下的简单标签：

```
<s:withAttribute name="SomeValue"/>
```

该标签的作用是将获得的请求参数 name 的值设置到标签的 name 属性，并在 JSP 页面输出标签属性值。例如，在浏览器地址栏输入

```
http://localhost:8080/mywebapp/index.jsp?name="Mike"
```

标签会将 Mike 设置到标签属性 name 中并在 JSP 页面输出 Mike。

14. 通过继承 SimpleTagSupport 创建一个如下的简单标签：

```
<s:modifybodycontent>bodyContent</s:modifybodycontent>
```

该标签的作用是在 JSP 页面输出标签处理器指定的内容 New body content，而不是原标签体内容 bodyContent。

第10章

JPA

Java 应用将数据持久化到关系数据库可以使用 JDBC 或 JPA(Java Persistent API)或其他第三方 O/R 映射框架。

10.1 JPA 简介

10.1.1 O/R 映射与 JPA

1. O/R 映射

Java 应用中运行的都是对象。将应用中的对象模型映射到关系模型的技术称为对象/关系映射(Object-Relational Mapping),简称 O/R 映射。借助于对象关系映射,可以实现对象与关系数据库数据之间的相互转换,解决面向对象与关系数据库存在的互不匹配问题。

Java 应用常用的 O/R 映射产品有 Hibernate、EclipseLink(以前的 TopLink)、MyBatis 等。这些 O/R 映射产品的出现早于 JPA 规范。

2. JPA

JCP 引入新的 JPA 规范出于两个原因:简化现有 Java EE 和 Java SE 应用开发工作;整合 ORM 技术以便提供 Java 统一的持久化标准。

JPA 规范主要包括如下内容:

(1) 提供标准的 O/R 映射。JPA 吸收了许多主流的持久化框架的优点,如 Hibernate、JDO 等。

(2) JPA 定义了用于持久化的实体对象,并推崇 POJO 的编程模型。

(3) JPA 没有同 Java EE 容器绑定在一起。这表示 JPA 可以在 EJB 3.0 中使用,还可以在 Web 应用或在脱离容器环境的 Java SE 中单独使用。

(4) 定义了服务提供者接口。在不用修改实体代码的前提下,可以使用不同的持久化提供者。例如,可以较容易地从 JPA 的实现产品 Hibernate 切换到 EclipseLink。

10.1.2 Eclipse 下搭建 JPA Java SE 环境

本节介绍在 Eclipse 下搭建 JPA Java SE 环境，实现使用 JPA 完成在数据库 user 表中增加一条记录。

1. 创建数据库和表

在 MySQL 5 数据库 testDB 中，表 user 的 DDL 如下：

```
CREATE TABLE `user` (
  `id` int(11) NOT NULL AUTO_INCREMENT,
  `name` varchar(255) DEFAULT NULL,
  `password` varchar(255) DEFAULT NULL,
  `age` int(11) DEFAULT NULL,
  `birthday` datetime DEFAULT NULL,
  PRIMARY KEY (`id`)
) ENGINE=InnoDB AUTO_INCREMENT=1 DEFAULT CHARSET=utf8;
```

2. 创建 Java 项目和配置构建路径

Eclipse 中创建 Java 项目 ch10-01 后，配置项目构建路径添加如下第三方类库：

- hibernate-release-4.3.10.Final\lib\required 下的所有 jar 文件；
- hibernate-release-4.3.10.Final\lib\jpa 下的 hibernate-entitymanager-4.3.10.Final.jar；
- MySQL\mysql-connector-java-5.0.6-bin.jar。

配置构建路径后项目 ch11-01 的结构如图 10-1 所示。

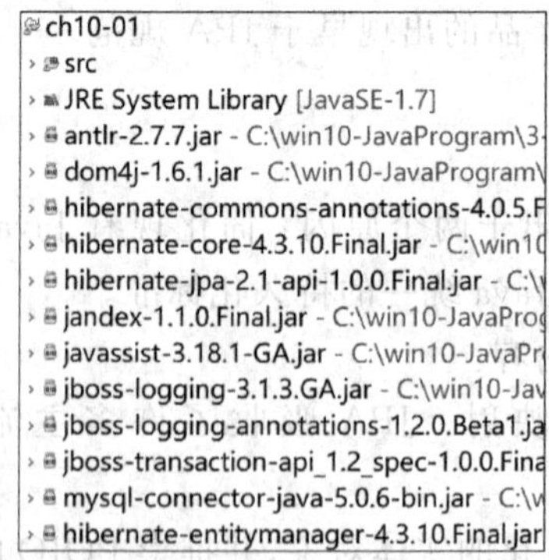

图 10-1 配置构建路径后项目 ch10-01 的结构

3. 创建 JPA 配置文件 persistence.xml

在项目 ch10-01 的 src 下创建文件夹 META-INF。在文件夹 META-INF 下创建

persistence. xml。persistence. xml 的代码如下：

```
<?xml version="1.0" encoding="UTF-8"?>
<persistence version="2.1"
    xmlns="http://xmlns.jcp.org/xml/ns/persistence"
    xmlns:xsi="http://www.w3.org/2001/XMLSchema-instance"
    xsi:schemaLocation="http://xmlns.jcp.org/xml/ns/persistence
    http://xmlns.jcp.org/xml/ns/persistence/persistence_2_1.xsd">
  <persistence-unit name="firstJPA2.1" transaction-type="RESOURCE_LOCAL">
     <class>entity.User</class>
        <properties>
           <property name="javax.persistence.jdbc.driver"
             value="com.mysql.jdbc.Driver" />
           <property name="javax.persistence.jdbc.url"
             value="jdbc:mysql://localhost:3306/testDB" />
           <property name="javax.persistence.jdbc.user" value="root" />
           <property name="javax.persistence.jdbc.password" value="root" />
           <property name="hibernate.dialect"
             value="org.hibernate.dialect.MySQL5Dialect" />
           <property name="hibernate.show_sql" value="true" />
           <property name="hibernate.hbm2ddl.auto" value="update" />
       </properties>
  </persistence-unit>
</persistence>
```

4. 编写实体类 User

在项目 ch10-01 的 src 下创建包 entity，并在 entity 下创建实体类 User。User 类的关键代码如下：

```
package entity;
//…导包略
@Entity
public class User {
    @Id
    private int id;
    private String name;
    private String password;
    private int age;
    private Date birthday;
    //…getter()和 setter()方法略
}
```

5. 编写测试类进行测试

编写测试类 UserTest，其关键代码如下：

```
package test;
…
public class UserTest {
    public static void main(String[] args) {
        EntityManagerFactory emfactory =
            Persistence.createEntityManagerFactory("firstJPA2.1");
        EntityManager em = emfactory.createEntityManager();
        em.getTransaction().begin();
        User user = new User();
        user.setName("张三");
        em.persist(user);
        em.getTransaction().commit();
        em.close();
        emfactory.close();
    }
}
```

运行 UserTest 类后,数据库 testDB 的 user 表增加了一条记录,如图 10-2 所示。

id	name	password	age	birthday
1	张三	(Null)	0	(Null)

图 10-2 运行 UserTest 类后数据库 testDB 的 user 表增加了一条记录

10.2 实 体

JPA 规范中的实体代表了关系数据库中的一张表,每个实体对象表示表的一行记录。实体对象以属性保存数据,实体对象的属性对应的是表的某一列。在基于 JPA 的应用中运行的是实体对象,需要编写对应的实体类。

10.2.1 实体类的编写规范

(1) 实体类前面必须使用@Entity 注解。

(2) 实体类必须有主键,用于唯一区分不同的实体;实体主键使用@Id 注解。

(3) 实体类必须有一个 public 或者 protected 的无参数的构造方法;实体类可以有其他构造方法。

(4) 实体类不能声明为 final 的;实体类的方法和属性都不能是 final 的。

(5) 实体对象如果需要被传送到远程客户,那么该实体类必须实现序列化接口 java.io.Serializable。

(6) 实体类或者非实体类都可以派生实体类;实体类可以派生非实体类。

(7) 在实体类中,持久化的实例变量必须是非 public 的(即或是 private 或是

protected 或是包访问权限)，并且只允许直接被实体类的方法访问；其他类只能通过实体类的 getter、setter 方法或者业务方法访问实体类实例变量。

10.2.2 @Entity 注解

@Entity 注解用于类前，指定该类是一个实体类。如下 User 类前使用@Entity 注解指定 User 类是实体类。

```
@Entity
public class User implements Serializable {
    ...
}
```

@Entity 唯一的 name 属性用于指定实体的名称。如下代码使用@Entity 的 name 属性指定实体的名称为 theUser。

```
@Entity(name="theUser")
public class User implements Serializable {
    ...
}
```

当不使用 name 属性时实体的名称取默认值。实体名称的默认值是实体类的非限定名(即不含包名的类名)。如下的@Entity 注解不使用 name 属性时实体的名称默认为类名 User。

```
@Entity
public class User implements Serializable {
...
}
```

在 JPA 查询语句中 from 关键字后面是实体名称，表示对实体进行查询。例如，如下的 User 实体使用 name 属性指定实体名称为 theUser 时，

```
@Entity(name="theUser")
public class User implements Serializable {
    ...
}
```

JPA 查询语句为：

```
select user from theUser user
```

User 实体不使用 name 属性指定实体名称时，

```
@Entity
public class User implements Serializable {
    ...
```

```
}
```

JPA 查询语句为：

```
select user from User user
```

10.2.3 @Table 注解

@Table 注解用于实体类的前面，其 name 属性指定了实体映射到的数据库表名。如下代码使用@Table(name="user")注解指定实体 User 映射到数据库的表名是 user。

```
@Entity
@Table(name ="user")
public class User implements Serializable {
    …
}
```

如果不使用@Table 注解，默认会使用实体类的类名作为映射到数据库的表名。如下代码不使用@Table 注解，User 实体映射到数据库的表名是类名 User。

```
@Entity
public class User implements Serializable {
    …
}
```

@Table 注解还有 catalog、schema、uniqueConstraints 属性，分别用于指定数据库的 catalog、映射表所属的 schema、唯一性字段约束。@UniqueConstraint 注解的 columnNames 属性指定唯一性字段的名称；如果唯一性字段由多个字段组成，就需要用{}，并在字段之间用逗号分隔。如下代码使用@UniqueConstraint 注解指定 id 和 name 字段为唯一性约束。

```
@Entity
@Table(name ="user",
    uniqueConstraints={@UniqueConstraint(columnNames={"id","name"})})
public class User implements Serializable {
    …
}
```

10.2.4 @Id 注解

@Id 注解用于实体类属性前面，指定该属性作为实体主键。如下代码使用@Id 注解指定属性 id 是实体 User 的主键。

```
public class User implements Serializable {
```

```
    @Id
    @GeneratedValue
    @Column(name ="id")
    private int id;
...
}
```

实体主键分为简单主键和主键类。简单主键的类型可以是 Java 基本类型(byte、int、short、long、char);或者是基本类型对应的包装类(Byte、Integer、Short、Long、Character);或者是 java. lang. String、java. util. Date、java. sql. Date、java. lang. BigInteger。float、double 及其包装类 Float、Double 与 java. math. BigDecimal 一般不建议用作主键类型。

@GeneratedValue 注解与@Id 一起使用,指定主键的生成方式。只有在持久化驱动生成数据库 schema 时才需要指定该注解;如果数据库及其表已经存在,那么编写实体类时可以不使用该注解。

@GeneratedValue 注解的 strategy 属性用于指定主键的生成方式。strategy 的取值可以是 GenerationType. SEQUENCE、GenerationType. IDENTITY、GenerationType. TABLE、GenerationType. AUTO 之一。

@GeneratedValue 注解的 generator 属性用于指定主键的生成器名称。

@GeneratedValue 等价于@GeneratedValue(strategy= GenerationType. AUTO),是主键生成的默认方式。此时指定持久化提供者决定主键生成方式,主键生成器使用 ID 生成器。

在一些复杂的场合,一个实体类 A 的主键是另外一个类 B 的对象的时候,类 B 称之为主键类。限于篇幅所限,本书不介绍主键类。

10. 2. 5 @Column 注解

@Column 注解用于实体类属性前,指定实体类属性与数据表中字段的映射关系。该注解定义了多个属性用于指定实体属性与数据库字段的映射信息。

- name 属性:指定映射到数据库表的字段名。默认值是属性名,即映射到数据库表的字段名为属性名。
- unique 属性:指定映射到数据库表的字段是否唯一。默认是 false,表示不唯一。
- nullable 属性:指定映射到数据库表的字段值是否允许为空。默认是 true,表示允许为空。
- insertable 属性:指定进行插入操作时生成的 insert 语句中该字段是否允许出现。默认是 true,表示允许出现。
- updatable 属性:指定进行更新操作时生成的 update 语句中该字段是否允许出现。默认是 true,表示允许出现。
- table 属性:指定映射到数据库表的字段所属的表。默认是实体类名,表示字段属于实体类类名所在的表。

- length 属性：指定映射到数据库表的字段的长度。默认是 255，仅适用于取值为字符串的字段。
- columnDefinition 属性：指定创建表时字段创建的 DDL。

如下代码使用@Column 注解的 name 属性指定 String name 属性映射到数据库表的 name 字段，该字段值允许为空，该字段长度是 50，该字段允许出现在 insert 和 update 语句中。

```
@Column(name ="name", nullable =false, length =50, insertable =true,
    updatable =true)
private String name;
```

如果在实体类属性前没有使用@Column 注解，那么该实体类中所有的没有使用@Transient 注解的属性将按照@Column 注解的默认属性值进行持久化。例如，实体类属性 String password 不使用@Column 注解时，该属性将按照@Column 注解的默认属性值进行持久化

```
private String password;
```

相当于以下代码：

```
@Column(name ="password", nullable =false, length =255, insertable =true,
    updatable =true)
private String password;
```

10.2.6 @Transient 注解

@Transient 注解使用在实体类的属性前，表示该属性不需要持久化。在保存或更新实体对象时，该属性不会持久化到关系数据库中。例如，如下 String password 属性前使用@Transient 注解指定该属性不需要持久化。

```
@Transient
private String password;
```

10.2.7 属性注解使用的位置

属性注解@Column、@Id、@Transient 等可以使用在实体类的属性之前，也可以使用在属性的 getter 方法之前。例如，如下的@Column 使用在属性 name 的 getter()方法之前。

```
@Column(name ="name", nullable =false, length =50, insertable =true,
    updatable =true)
public String getName() {
```

```
        return name;
    }
```

一个实体类中不能在属性和该属性的 getter()方法之前同时使用@Column 注解。例如，如下在 name 属性及其 getter 方法之前同时使用@Column 是错误用法。

```
@Column(name ="name", nullable =false, length =50, insertable =true,
    updatable =true)
private String name;
…
@Column(name ="name", nullable =false, length =50, insertable =true,
    updatable =true)
public String getName() {
    return name;
}
```

10.3 EntityManager

持久化上下文是内存中实体对象与数据库之间的连接枢纽。某个实体对象只有位于持久化上下文中时才与数据库中的记录对应。EntityManager 接口的实例可以访问持久化上下文，实现创建、删除实体对象，通过实体主键查找实体，对实体进行查询等操作。

10.3.1 获取 EntityManager 实例

EntityManager 对象的管理方式有两种：容器管理的 EntityManager 对象和应用管理的 EntityManager 对象。

对于容器管理的 EntityManager 对象，程序员不需要考虑 EntityManager 对象的连接和释放以及事务等复杂的问题；这些都由容器管理。容器管理的 EntityManager 对象必须在 EJB 容器中运行，而不能在 Web 容器和 J2SE 的环境中运行。如果使用 EJB 尽可能选择容器管理的 EntityManager 对象。

对于应用管理的 EntityManager 对象，程序员需要手动地控制 EntityManager 对象的连接和释放、手动地控制事务等。但这种方式获得的 EntityManager 对象不仅可以在 EJB 容器中使用，也可以在脱离 EJB 容器的任何的 Java 环境使用，例如，在 Web 容器、Java SE 环境中使用应用管理的 EntityManager。

1. 获取 EJB 容器管理的 EntityManager

通过在 EntityManager 对象前使用注解@PersistenceContext 实现获取 EJB 容器管理的 EntityManager 对象。

如下代码演示了 UserDAOBean 获取 EJB 容器管理的 EntityManager。在 EJB 3 中，

通常一个会话 Bean 组件会有远程接口(或本地接口)及其实现类。本例中远程接口是 UserDAO,其实现类是 UserDAOBean。

```
public interface UserDAO {
    //新增一个 User
    public void add(User user);
}

@Stateless
@Remote
public class UserDAOBean implements UserDAO {
    @PersistenceContext(unitName="firstJPA")
    private EntityManager em;
    public void add(User user) {
        em.persist(user);
    }
}
```

在 UserDAOBean 中获取 EJB 容器管理的 EntityManager 对象 em 可通过如下代码完成:

```
@PersistenceContext(unitName="firstJPA")
private EntityManager em;
```

firstJPA 是持久化单元名称。UserDAOBean 的 add()方法使用 EntityManager 接口定义的 persist()方法,通过保存 user 对象实现增加一条记录的操作。

2. 获取应用管理的 EntityManager

获得应用管理的 EntityManager 对象,首先需要使用

```
Persistence.createEntityManagerFactory()
```

方法获得 EntityManagerFactory 对象;再使用 EntityManagerFactory 对象的 createEntityManager()方法创建 EntityManager 对象。如下代码演示了在 Java SE 环境下获取应用管理的 EntityManager 对象。

```
public class UserTest {
    public static void main(String[] args) {
        EntityManagerFactory emfactory =
            Persistence.createEntityManagerFactory("firstJPA2.1");
        EntityManager em =emfactory.createEntityManager();
        em.getTransaction().begin();
        ...
        em.persist(user);
        ...
    }
```

```
}
```

上述 Java 应用程序在其 main()方法中通过

```
Persistence.createEntityManagerFactory("firstJPA2.1")
```

获得一个 EntityManagerFactory 对象，firstJPA2.1 是持久化单元名称；接着通过 emfactory.createEntityManager()获得 EntityManager。

10.3.2 配置持久化单元

1. 持久化单元与 persistence.xml

持久化单元维护一组实体类、映射元数据、数据库相关的配置数据。JPA 需要在配置文件 persistence.xml 中配置持久化单元。如果应用中没有 persistence.xml，就无法访问持久化单元，进而无法获取和使用访问实体的 EntityManager。JPA 要求 persistence.xml 位于 ejb-jar、ear、war、jar 文件的 META-INF 目录中。

2. 配置持久化单元

在 persistence.xml 中，<persistence-unit name="firstJPA2.1">配置了持久化单元的名称为 firstJPA2.1。

persistence.xml 文件中可以配置多个持久化单元。如下 persistence.xml 中配置了两个持久化单元 firstJPA2.1 和 otherPersinstenceUnit。

```
<?xml version="1.0" encoding="UTF-8"?>
<persistence version="2.1"
    xmlns=http://xmlns.jcp.org/xml/ns/persistence…>
    <persistence-unit name="firstJPA2.1" transaction-type="RESOURCE_LOCAL">
        …
    </persistence-unit>
    <persistence-unit name="otherPersinstenceUnit">
        …
    </persistence-unit>
</persistence>
```

3. EJB 容器管理的 EntityManager 指定持久化单元

EJB 容器管理的 EntityManager 显式指定持久化单元如下：

```
@PersistenceContext("持久化单元名称")
private EntityManager em;
```

如果 persistence.xml 中只有一个持久化单元，在 EJB 容器管理的 EntityManager 中则可以不用显式指定持久化单元。例如，只有一个持久化单元时，EJB 容器管理的 EntityManager 不需要没有显式指定持久化单元的代码如下：

```
@PersistenceContext
private EntityManager em;
```

4. 应用管理的 EntityManager 指定持久化单元

在获取应用管理的 EntityManager 中,必须显式指定持久化单元。例如,如下代码在获取应用管理的 EntityManager 中显式指定持久化单元为 firstJPA2.1。

```
Persistence.createEntityManagerFactory("firstJPA2.1");
```

5. persistence.xml 的其他常用元素说明

- <provider>:指定一个持久化提供者(可选)。
- <transaction-type>:指定事务类型。其值可以是 RESOURCE_LOCAL 或 JTA。
- <jta-data-source>、<non-jta-data-source>:可选。用于分别指定持久化提供者使用的 JTA 或 non-JTA 数据源的全局 JNDI 名称。
- <mapping-file>:可选。指定以 XML 方式配置实体类的映射文件路径。
- <jar-file>:可选。指定持久化单元中其他的 jar 文件的路径。该路径以包含 persistence.xml 的 jar 文件为基准。
- <class>:指定持久化单元中需要使用的实体类。在 Java SE 环境中应该显式列出(可选)。
- <exclude-unlisted-classes>:可选。指定持久化单元中不需要使用的实体类。
- <properties>:厂商专有属性。Java SE 环境下通常配置数据库连接信息。

10.3.3 实体对象的状态与 EntityManager API

1. 实体对象的状态

一个实体对象具有持久化身份是指该实体对象与数据库中的一条记录关联。根据实体对象是否具有持久化身份和实体对象是否处于持久化上下文中,实体对象在其生命周期中的状态分为 4 种。

1) 新建状态

若实体对象没有持久化身份,并且也没有处于持久化上下文中,则称该实体对象此时处于新建状态。处于新建状态的实体对象的状态的改变不会同步到数据库中。

2) 托管状态

若实体对象具有持久化身份,并处于持久化上下文中,则称该实体对象此时处于托管状态。

3) 游离状态

若实体对象具有持久化身份,但是并不处于持久化上下文中,此时称该实体对象处于游离状态。一个托管状态的实体对象在持久化上下文结束后处于游离状态。

4）删除状态

实体对象处于持久化上下文中，但是此时数据库不存在对应的记录。

实体对象的状态变化在事务提交或者显式调用 EntityManager 的 flush()方法后会同步到数据库中。

2. 调用 EntityManager API 的方法导致实体对象状态变迁

调用 EntityManager API 中的方法导致实体对象状态变迁如图 10-3 所示。

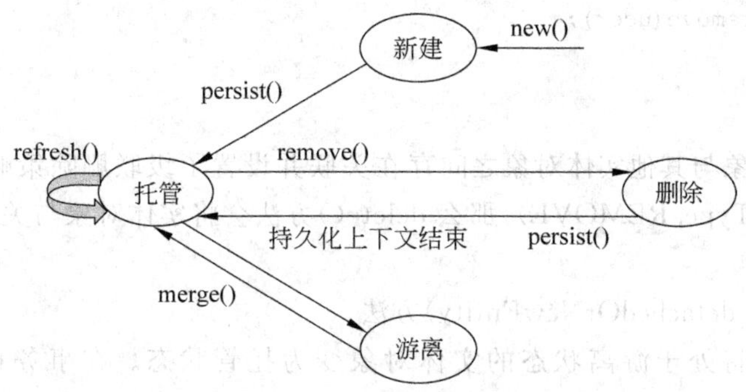

图 10-3　调用 EntityManager API 中的方法导致实体对象状态变迁

在使用 new 关键字创建一个实体对象后，到调用 EntityManager 的 persist()方法之前，该实体对象此时处于新建状态。例如，如下 add(String name，String password)方法中，从 User user = new User()到 user. setPassword(password)这 3 行代码，user 对象处于新建状态。

```
public User add(String name,String password) {
    User user =new User();
    user.setName(name);
    user.setPassword(password);
    em.persist(user);
    user.setName("Java EE");
    return user;
}
```

1）persist(Object newEntity)方法

该方法用于将一个新建状态的实体对象变成托管状态。在事务提交或显式调用 flush()方法后，数据库会新增一条记录。例如，add(String name，String password)方法中，在执行 em. persist(user)后，实体对象 user 由新建状态变成托管状态；接着调用 user. setName("Java EE")，处于托管状态的 user 实体对象的 name 的状态变化在事务提交或者显式调用 EntityManager 的 flush()方法后会同步到数据库。

如果实体对象与其他实体存在关联关系并且设置了级联保存策略(CascadeType. ALL 或 CascadeType. PERSIST)，那么 persist()方法会同时将实体对象与关联实体对象保存到数据库。

2) remove(Object managedEntity)方法

该方法用于将一个处于托管状态的实体对象变为删除状态。在事务提交或显式调用flush()方法后，实体对象对应的数据库记录将被删除。例如，如下 delete()方法中调用em. remove(user)后，user 实体对象由托管状态变为删除状态。

```
public void delete(int id) {
    User user =em.find(User.class, id);
    if (user !=null) {
        em.remove(user);
    }
}
```

如果实体对象与其他实体对象之间存在关联并设置了级联删除策略(CascadeType. ALL 或 CascadeType. REMOVE)，那么 delete()方法会将实体对象与关联实体对象都变为删除状态。

3) merge(T detachedOrNewEntity)方法

该方法用于将处于游离状态的实体对象变为托管状态。在事务提交或显式调用flush()方法后，实体对象的状态变化会同步到数据库。例如，如下 update()方法中调用em. merge(detachedUser)后，处于游离状态的实体对象 detachedUser 变为托管状态。

```
public void update(User detachedUser) {
    User managedUser =em.merge(detachedUser);
}
```

如果实体对象与其他实体对象之间存在关联并设置了级联合并策略(CascadeType. ALL 或 CascadeType. MERGE)，那么 merge()方法会将实体对象与关联实体对象的变化反映到数据库。

4) find(Class<T> entityClass, Object primaryKey)方法

该方法用于根据传入的实体类类型和实体对象主键来获得实体对象。例如，如下的get()方法中，em. find(User. class, id)根据传入的 id 和 User. class 获得一个实体对象 user。

```
public User get(int id) {
    User user =em.find(User.class, id);
    return user;
}
```

持久化提供者使用缓存机制，托管的实体可能位于缓存中。当查找实体时可能不需要从数据库获得数据。find()方法首先从缓存中查找到的实体，如果找到就返回该实体；否则会创建一个新的实体并从数据库加载其持久状态(这称为即时加载)。如果数据库不存在实体主键指定的记录，则 find()方法返回 null。

5) getReference(Class<T> entityClass, Object primaryKey)方法

该方法用于根据传入的参数实体类和实体主键来查找实体。getReference()方法首

先从缓存中查找实体对象,如果找到就返回该实体;否则会创建一个新的实体,但不会立即从数据库加载其持久状态。在第一次访问该实体对象的某个持久化属性时才从数据库加载其持久状态(这被称为延迟加载。延迟加载机制可以避免从数据库加载持久化状态的性能开销)。getReference()方法查找实体时,如果找不到主键对应的数据库记录,不会返回 null 而是抛出异常 javax. persistent. EntityNotFoundException;该异常不是在调用 getReference()方法时抛出,而是在第一次访问实体属性时抛出。

10.3.4 刷新操作

1. refresh(Object entity)方法

该方法用于将实体对应的数据库表的记录重新加载到实体的属性中。例如,如下代码使用 refresh(user)方法将数据库表的记录加载到 user 实体对象。

```
public User get(int id) {
    User user =em.find(User.class, id);
    …
    em.refresh(user);               //将 user 表中记录设置到 user 对象的属性中
    return user;
}
```

2. flush()方法

flust()方法用于将持久化上下文中的所有实体状态同步到数据库中。

在调用 EntityManager 的 persist()、merge()、remove()或者托管对象的 setter()方法时,实体的状态并不会立刻同步到数据库中;而是由 JPA 缓存起来,在执行 flush()时写入。在事务提交的时候,JPA 会自动执行 flush()一次性保存所有数据。如果需要立即保存,可手动执行 flush()。

例如,如下代码调用 flush()方法将所有实体状态变化同步到数据库。

```
@PersistContext
private EntityManager em;
…
    public void update2(User detachedUser) {
        em.setFlushMode(FlushModeType.COMMIT);
        User managedUser =em.merge(detachedUser);
        managedUser.setName("JavaEE 5");
        String sql ="select user from User user where user.id=?1";
        Query query =em.createQuery(sql);
        query.setParameter(1, new Integer(1));
        query.executeUpdate();
        em.flush();                 //进行刷新操作
        …                           //其他更新操作
```

```
    }
...
```

3. setFlushMode(FlushModeType flushMode)方法

该方法用于设置持久化上下文的刷新模式。

刷新模式有两种：FlushModeType. AUTO 和 FlushModeType. COMMIT，默认是 AUTO，表示刷新是使用 javax. persistence. Query 对象执行查询之前或者事务提交时才进行。COMMIT 模式表示只有在事务提交时才进行刷新操作。对于容器管理事务，事务提交的时间是在业务方法结束之前。

find()、getReference()进行的查询都不会引起刷新，因为通过主键查询，实体不会受到任何更新操作的影响。

如下代码使用 setFlushMode()方法设置刷新模式为 FlushModeType. COMMIT，刷新操作在 update2 方法结束之前进行。

```
@PersistContext
private EntityManager em;
...
    public void update2(User detachedUser) {
        em.setFlushMode(FlushModeType.COMMIT);
        User managedUser =em.merge(detachedUser);
        managedUser.setName("Java Web");
        String sql ="select user from User user where user.id=?1";
        Query query =em.createQuery(sql);
        query.setParameter(1, new Integer(1));
        query.executeUpdate();
        …//其他更新操作
    }
...
```

10.3.5 实体生命周期回调

JPA 为实体定义的生命周期事件包括 PrePersist、PostPersist、PreUpdate、PostUpdate、PreRemove、PostRemove、PostLoad。在调用 persist()、merge()、remove()、find()方法或者执行 JPQL 查询时，会触发实体的生命周期事件。例如，persist()触发数据插入事件，merge()方法触发数据更新事件，remove()方法触发数据删除事件，JPQL 查询实体触发数据加载事件。持久化提供者得到这些事件通知后，转而调用实体类中定义的事件处理方法或者调用一个外部类中的事件处理方法。

1. 在实体类中定义生命周期回调方法

在实体类中编写一个方法后使用 JPA 提供的相应的注解声明该方法为生命周期回

调方法。

例如,如下的 User 实体类在 beforePersist()方法前使用注解@PrePersist,声明该方法在 persist()之前被调用。

```
@Entity
public class User implements Serializable {
    ...
    @PrePersist
    void beforePersist(){
        System.out.println("before User Entity be Persisted");
    }
    ...
}
```

其他的注释如@PostPersist、@PostUpdate 等用法与此类似。

2. 使用监听器类定义生命周期回调方法

使用监听器类定义生命周期回调方法需要如下两个步骤。

1) 定义监听器类

在监听器类的方法之前使用 JPA 的相应注解声明该方法为生命周期回调方法。

例如,如下代码定义了监听器类 UserListener,并在其 beforePersist(User user)方法之前使用注解@PrePersist,声明该方法在实体 User 持久化之前被调用。

```
public class UserListener {
    ...
    @PrePersist
    void beforePersist(User user){
        System.out.println("before User Entity be Persisted");
    }
    ...
}
```

2) 在实体类中声明监听器类

在实体类之前使用注解@EntityListeners({实体类类名. class})指定实体类的监听器类。

例如,如下代码在 User 实体类之前使用@EntityListeners({UserListener. class})指定 User 实体的监听器类是 UserListener. class。

```
@Entity
@EntityListeners({UserListener.class})
public class User implements Serializable {
    ...
}
```

如果 User 实体有多个监听器类,可以在每个监听器类之间加“,”号进行分隔。

例如，如下代码使用@EntityListeners({UserListener.class，AnotherListener.class})指定User实体有两个监听器类：UserListener.class和AnotherListener.class。

```
@EntityListeners({UserListener.class,AnotherListener.class})
public class User implements Serializable {
...
}
```

10.4 实体映射关系

关系数据库中有4种关系：一对一、一对多、多对一、多对多。关系数据库中的关系没有方向性。

JPA中的关系存在方向性。关系的方向性分单向和双向。在单向关系中，只有关系的拥有方才知道对方；在双向关系中，关系的双方都知道对方的存在。

由于JPA双向一对多和双向多对一是相同的，因此JPA总共有7种关系：单向一对一、双向一对一、单向一对多、双向一对多、单向多对一、单向多对多、双向多对多。

10.4.1 单向一对一映射

单向一对一关系是指一个实体A与另一个实体B之间存在关系；通过实体A可以获得实体B，通过实体B不能获得实体A。

关系数据库一对一关系的表设计有3种方式：外键关联、关联表、共享主键。最常见的是外键关联方式。在单向一对一映射中采用外键关联方式。

人和身份证作为一个单向一对一的关系，通过人可以获得其身份证；通过身份证不需要获得其主人。

在关系数据库中，人和身份证的一对一关系采用外键关联方式的数据库表如图10-4所示。

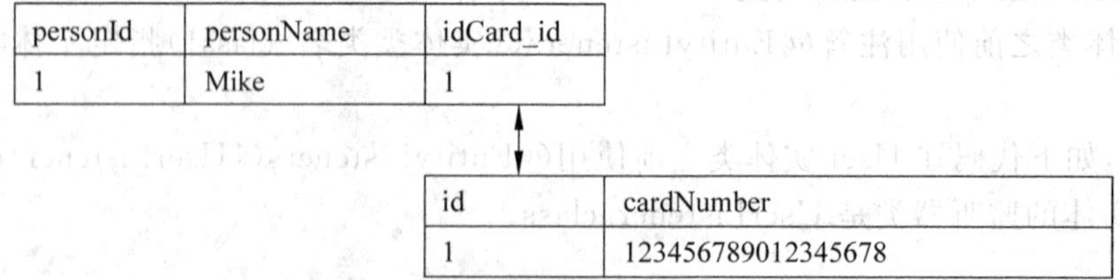

personId	personName	idCard_id
1	Mike	1

id	cardNumber
1	123456789012345678

图10-4 人和身份证的一对一关系采用外键关联方式的数据库表

为了将图10-4所示的表映射为JPA的单向一对一关系，需要使用Java代码编写实体Person和IDCard。Person.java的关键代码如下：

```
package entity;
```

```
...
@Entity
@Table(name ="person")
public class Person implements Serializable {
    @Id
    @GeneratedValue
    @Column(name ="personId")
    private int personId;
    @Column(name ="personName", nullable =false, length =50, insertable =true,
        updatable =true)
    private String personName;
    @OneToOne(optional =true, cascade =CascadeType.ALL)
    @JoinColumn(name ="idCard_id", unique =true)
    private IDCard icCard;
    ...
}
```

IDCard.java 的关键代码如下：

```
package entity;
...
@Entity
public class IDCard implements Serializable {
    @Id
    @GeneratedValue
    private int id;
    @Column(nullable =false, length =18, unique =true)
    private String cardNumber;
    ...
}
```

单向一对一映射，需要在关系拥有方添加关系相对方的属性及其 getter()和 setter()方法，并在该属性(或属性的 getter()方法)前使用@OneToOne 注解。关系相对方不用定义关系拥有方的属性。

在 Person 和 IDCard 实体关系中，Person 实体是关系拥有方，IDCard 实体是关系相对方。于是在 Person 类中添加了 IDCard 属性及其 getter()和 setter()方法，并在该属性前使用@OneToOne 注解。

@OneToOne 注释常用属性说明如下。

- fetch：指定关联实体的加载方式。取值有两种，分别是 FetchType.EAGER(立即加载)和 FetchType.LAZY(延迟加载)。立即加载表示在加载实体的同时加载关联实体的属性。例如，将本例的 Person 实体@OneToOne 注释使用 fetch = FetchType.EAGER 设置了立即加载，那么在加载 Person 实体时会同时加载关联的实体属性 IDCard。在一对一关系中，fetch 属性的默认值是 FetchType.

EAGER。

- cascade：定义关联实体之间的级联关系。cascade 的取值可以是 CascadeType. PERSIST（级联保存）、CascadeType. REMOVE（级联删除）、CascadeType. REFRESH（级联刷新）、CascadeType. MERGE（级联更新）这 4 个中的一个或者多个；如果想同时包括这 4 项，需要将 cascade 的取值设为 CascadeType. ALL。根据不同的级联关系，对当前实体的操作会影响到关联的实体。例如，假如 Person 与 IDCard 之间是级联删除关系，那么删除 Person 会同时删除 IDCard 对象。如果 IDCard 和其他实体还有级联关系，那么这样的操作会递归进行。
- optional：指定一对一关系中关系相对方是否可以为空，其取值有 true 和 false。如果取值为 false，则关系相对方不能为空；如果取值为 true，则关系相对方可以为空。optional 的默认值是 true。
- targetEntity：用于指定需要关联的实体类。一般不需要指定该属性，持久化提供者会根据关联的实体自动配置该属性。
- @JoinColumn：指定一对一关系中关系拥有方的数据库表的外键。@JoinColumn 注释中的 name = "idCard_id"指定了外键名称是 idCard_id；unique = true 指定该外键的值具有唯一性，不能重复。@JoinColumn 不是必需的。如果没有该注释，持久化提供者会自动生成外键映射，外键的命名规则是映射属性名_关系相对方表的主键名。此处 idCard_id 外键名与不使用@JoinColumn 注释生成的外键名一样。

测试类 UniPersonIDCardTest 来完成实体 Person 的持久化，其关键代码如下：

```
public class UniPersonIDCardTest {
    public static void main(String[] args) {
        EntityManagerFactory emfactory =
            Persistence.createEntityManagerFactory("firstJPA2.1");
        EntityManager em =emfactory.createEntityManager();
        em.getTransaction().begin();
        Person person =new Person();
        person.setPersonName("Mike");
        IDCard idCard =new IDCard();
        idCard.setCardNumber("123456789012345678");
        person.setIdCard(idCard);
        em.persist(person);
        em.getTransaction().commit();
        em.close();
        emfactory.close();
    }
}
```

上述代码首先创建 Person 对象和 IDCard 对象，然后调用

```
person.setIdCard(idCard);
```

方法设置 person 中的 idCard 对象，再调用 persist(person)方法。由于 Person 类中的级联操作为 CascadeType. ALL，这导致 person 和 idCard 对象同时持久化到数据库。

10.4.2 双向一对一映射

双向一对一关系是指两个实体之间存在关联，从实体的任何一方能够获得另一方。

人和身份证构成双向一对一关系：每个人有唯一的身份证；每张身份证只属于唯一的人。通过人可以获取其身份证，通过身份证可以获取身份证的主人。

双向一对一映射的数据库表采用基于外键关联，这与 10.4.1 节的单向一对一映射对应的数据库表相同，如图 10-4 所示。

双向一对一映射的实体类 Person. java 与 10.4.1 节的单向一对一 Person 完全相同。

IDCard. java 的关键代码如下：

```
package entity;
...
@Entity
public class IDCard implements Serializable {
    @Id
    @GeneratedValue
    private int id;
    @Column(nullable=false,length=18,unique=true)
    private String cardNumber;
    @OneToOne(optional=false,cascade=CascadeType.REFRESH,mappedBy="idCard")
    private Person person;
    ...
}
```

IDCard 中需要增加 Person 的属性及其 getter 和 setter 方法，并使用@OneToOne 注解，指定 IDCard 与 Person 具有一对一关系。

mappedBy 属性指定实体之间双向关系的关系维护端，即指定双向关系由哪一方来维护关系。IDCard 类中 mappedBy＝"idCard"表示关系是由 idCard 所属的类 Person 维护关系，负责数据库中的 person 表的外键 idCard_id 的更新。

mappedBy 属性只有在双向关系@OneToOne、@OneToMany、@ManyToMany 中才能使用；如果实体之间是单向关系，则不需要在@OneToOne、@OneToMany、@ManyToMany上使用 mappedBy 属性。

optional＝false 指定关联的 person 不能为 null。

cascade 属性指定到目标实体 person 的级联操作类型为级联刷新。

10.4.3 单向一对多映射

现实应用中一对多关系更为常见。关系数据库中表示一对多关系有两种方式：中间表方式和外键关联方式。在此介绍使用基于中间表方式实现实体单向一对多映射，这也是 JPA 一对多映射的默认方式。

部门与员工是一对多关系，一个部门可以拥有多个员工，每个员工属于一个部门。

采用中间表方式的部门与员工表如图 10-5 所示。中间表的两列是外键：第一个外键(department_id)是关系一方主键；第二个外键(employee_id)是关系多方的外键。

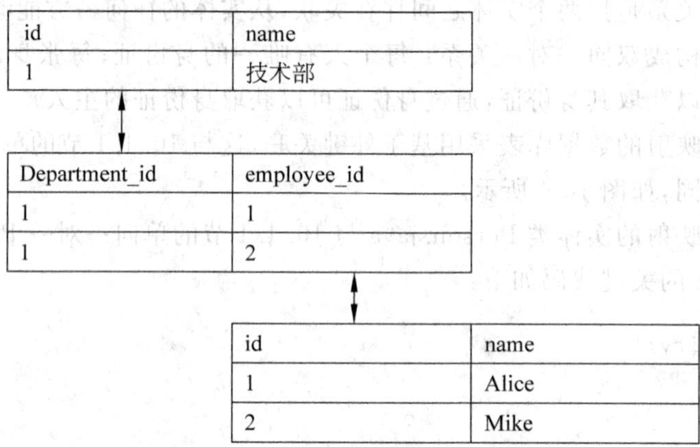

id	name
1	技术部

Department_id	employee_id
1	1
1	2

id	name
1	Alice
2	Mike

图 10-5　采用中间表方式的部门与员工表

Department 的关键代码如下。

```
package entity;
...
@Entity
public class Department implements Serializable {
    @Id
    @GeneratedValue
    private int id;
    private String name;
    @OneToMany(cascade =CascadeType.ALL, fetch =FetchType.EAGER)
    @JoinTable(name ="department_employee",
        joinColumns ={
            @JoinColumn(name ="department_id",
                referencedColumnName ="id") },
        inverseJoinColumns ={
            @JoinColumn(name ="employee_id",
                referencedColumnName ="id") })
    private Collection<Employee>employees;
    ...
}
```

Employee 的关键代码如下：

```
package entity;
...
```

```
@Entity
public class Employee implements Serializable {
    @Id
    private int id;
    private String name;
    …
}
```

在单向一对多的映射中，需要在关系拥有方添加关系相对方的属性及其 getter 和 setter 方法，并在该属性上使用@OneToMany 注解。关系相对方不用定义关系拥有方的属性。在 Department 和 Employee 实体关系中，Department 实体是关系拥有方，Employee 实体是关系相对方。于是在 Department 类中添加了 Collection<Employee> employees 及其 getter 和 setter 方法，并使用@OneToMany 注解指定 Department 和 Employee 是一对多关系。

@JoinTable 注解指关联表，其 name 属性指定关联表表名，此处关联表名为 department_employee。

@JoinTable 注解的 joinColumns 属性指定中间表中一方外键。其@JoinColumn 注解的 name 属性指定中间表中一方的外键名称；referencedColumnName 指定 department_id 外键参考了 department 表的 id 主键。

@JoinTable 注解的 inverseJoinColumns 属性指定中间表中多方外键。其@JoinColumn 注解的 name 属性指定中间表中多方的外键名称为 employee_id；referencedColumnName 指定 employee_id 外键参考了 employee 表的 id 主键。

10.4.4 双向一对多映射

在双向一对多关系中，从关系中的一方可以获得多方；反之，从关系中的多方也可以获得一方。

10.4.3 节介绍了数据库一对多关系中基于中间表方式实现 JPA 实体映射。数据库一对多关系用得最多的是外键关联，即在多方设置外键指向一方主键。

部门与员工是一对多关系，一个部门可以拥有多个员工，每个员工属于一个部门。

采用外键关联方式的部门与员工表如图 10-6 所示。

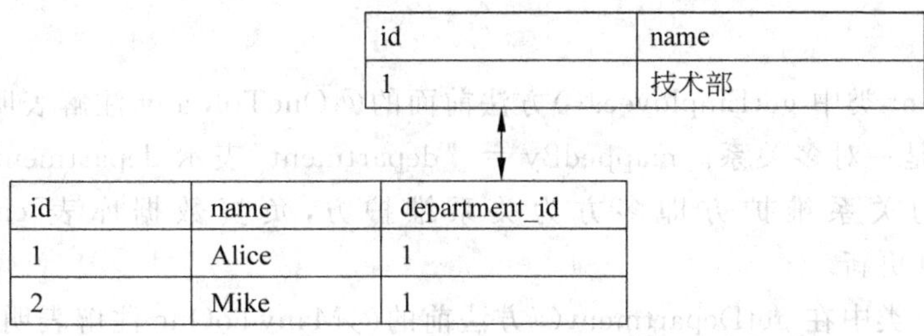

id	name
1	技术部

id	name	department_id
1	Alice	1
2	Mike	1

图 10-6 采用外键关联方式的部门与员工表

基于外键关联的部门实体类 Department 的关键代码如下：

```
package entity;
...
@Entity
public class Department implements Serializable {
    @Id
    private int id;
    private String name;
    @OneToMany(cascade =CascadeType.ALL, fetch =FetchType.EAGER,
        mappedBy ="department")
    private Collection<Employee>employees;
    ...
}
```

员工实体类 Employee 的关键代码如下：

```
package entity;
...
@Entity
public class Employee implements Serializable {
    private int id;
    private String name;
    private Department department;
    @Id
    @GeneratedValue
    public int getId() {
        return id;
    }
    ...
    @ManyToOne
    @JoinColumn(name="department_id")
    public Department getDepartment() {
        return department;
    }
    ...
}
```

Department 类中 getEmployees()方法前面的@OneToMany 注解表明 Department 与 Employee 是一对多关系。mappedBy ＝ "department"表示 department 属性所在的 Employee 类为关系维护方即多方为关系维护方，负责数据库表 employee 外键 department_id 更新。

Employee 类中在 getDepartment()方法前的@ManyToOne 注解表明 Employee 与 Department 是多对一关系；@JoinColumn(name＝" department_id")注解指定外键名称为 department_id。

10.4.5 单向多对一映射

在一个多对一关系中，从多方可以获得一方；从一方不需要获得多方。

员工与部门是多对一的关系。从员工可以获得部门，从部门不需要获得员工。这是一个单向多对一关系。采用外键关联方式的员工与部门的数据库表见图 10-4。

员工实体类 Employee 的关键代码如下：

```
package entity;
…
@Entity
public class Employee implements Serializable {
    @Id
    private int id;
    private String name;
    @ManyToOne(cascade =CascadeType.ALL)
    private Department department;
    …
}
```

关系相对方是部门实体 Department，其关键代码如下：

```
package entity;
…
@Entity
public class Department implements Serializable {
    private int id;
    private String name;
    @Id
    public int getId() {
        return id;
    }
    …
}
```

在单向多对一关系中，仅仅在关系拥有方 Employee 中使用了@ManyToOne 注解表明员工与部门是多对一的关系。

10.4.6 单向多对多映射

在一个 A 与 B 的多对多关系中，从多方 A 可以获得另外的多方 B；从 B 不需要获得 A。多对多映射时数据库需要采用中间表方式。

学生与课程是多对多的关系。从学生可以获得课程，从课程不用获得学生，这是一个单向多对多关系。学生与课程多对多关系的数据库表如图 10-7 所示。

关系拥有方学生实体 Student 的关键代码如下：

```
package entity;
...
@Entity
public class Student {
    @Id
    private int id;
    private String name;
    @ManyToMany(cascade=CascadeType.ALL,fetch=FetchType.EAGER)
    @JoinTable(name="student_course")
    private Collection<Course>courses =new ArrayList<Course>();
    ...
}
```

id	name
1	李明
2	张晓

student_id	course_id
1	Alice
2	Mike

id	courseName
1	数据库

图 10-7　学生与课程多对多关系的数据库表

在 getCourses()方法前使用@ManyToMany 注解表明 Student 与 Course 是多对多关系。@JoinTable(name="student_course")指定了中间表名称为 student_course。

关系相对方是课程。课程实体 Course 的关键代码如下：

```
package entity;
...
@Entity
public class Course {
    @Id
    private int id;
    private String courseName;
    ...
}
```

由于是学生和课程是单向多对多关系，故不需要在 Course 类使用多对多注解。

10.4.7 双向多对多映射

双向多对多关系是指两个实体之间存在多对多关联，从实体的任何一方都能够获得另一方。

学生与课程是多对多的关系。从学生可以获得课程；从课程也可以获得学生，这是一个双向多对多关系。学生与课程多对多关系的数据库表与单向多对多是相同的，见图 10-7。

学生实体类 Student 与单向多对多映射的 Student 完全相同。

课程实体类 Course 的关键代码如下：

```
package entity;
...

@Entity
public class Course {
    @Id
    private int id;
    private String courseName;
    @ManyToMany(mappedBy ="courses", cascade =CascadeType.ALL,
        fetch =FetchType.EAGER)
    private Collection<Student> students =new ArrayList<Student>();
    ...

}
```

在 Course 实体中增加了 Collection<Student> students 属性及其 getter()和 setter()方法。在 getStudents()方法前使用了@ManyToMany 注解。mappedBy 属性指定属性 courses 所在的类 Student 负责关系维护。

10.5 实体映射继承与多态

JPA 支持面向对象中继承的映射，即将应用中具有继承关系的实体类映射到关系数据库中的表。JPA 映射继承的方式有 3 种：整个类继承结构使用单个表；各子类使用单独的表；各个具体类使用单独的表。

例如，有一个由实体类 Person、Student、Teacher 组成的继承层次结构，如图 10-8 所示。Person 是整个继承层次结构的根，具有 name、sex、age 属性，充当主键的 id 属性未在图中列出；Student 和 Teacher 继承自 Person，分别具有 scholarship(奖学金)和 salary(工资)属性。对于如图 10-8 所示的继承层次结构，下面分别使用 JPA 的 3 种映射方式对其进行映射。

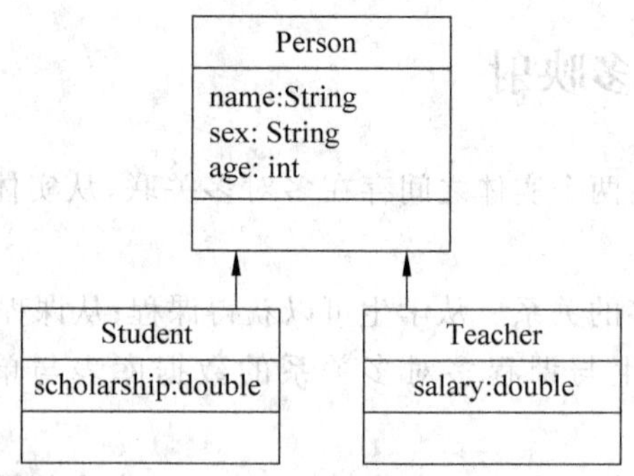

图 10-8　实体类 Person、Student、Teacher 组成的继承层次结构

10.5.1　整个类继承层次结构使用单个表

所有实体类的属性映射到一张表中，关系数据库中的表中提供单独的一列来区分继承层次结构中的不同类型。

Person、Student、Teacher 继承层次结构采用整个类继承层次结构使用单个表映射方式，其对应的数据库表 person 的 DDL 如下：

```
CREATE TABLE `person` (
  `identity` varchar(31) NOT NULL,
  `id` int(11) NOT NULL auto_increment,
  `name` varchar(255) default NULL,
  `sex` varchar(255) default NULL,
  `age` int(11) NOT NULL,
  `scholarship` double default NULL,
  `salary` double default NULL,
   PRIMARY KEY  (`id`)
) ENGINE=InnoDB DEFAULT CHARSET=utf8;
```

整个类继承层次结构中的所有属性在该表都能找到对应的字段。字段 id、name、sex、age 来自于父类 Person；字段 scholarship 来自于子类 Student；字段 salary 来自于子类 Teacher。该表增加的列 identity 用于区分继承层次结构中的不同类型，需要在父类 Person 中使用注解@DiscriminatorColumn 定义。

采用整个类继承层次结构使用单个表映射方式的 Person 类的关键代码如下：

```
package entity.inherit.singletable;
...
@Entity
@Inheritance(strategy=InheritanceType.SINGLE_TABLE)
@DiscriminatorColumn(name="identity",discriminatorType=DiscriminatorType.
STRING)
@DiscriminatorValue("PERSON")
```

```
public class Person implements Serializable{
    @Id
    protected int id;
    protected String name;
    protected String sex;
    protected int age;
    …
    public String toString(){
        return "id="+id+", name="+name+", sex="+sex+", age="+age;
    }
}
```

在最上层的根类中使用：

@Inheritance(strategy=InheritanceType.SINGLE_TABLE)指定映射继承的方式为整个类继承层次结构使用单个表；

@DiscriminatorColumn(name="identity",discriminatorType=DiscriminatorType.STRING)指定表中用于区分列的列名和该列的 Java 类型分别为 identity 和 String；

@DiscriminatorValue("PERSON")指定当前类在关系数据库表中区分列的值为 PERSON。

采用整个类继承层次结构使用单个表映射方式的子类 Student 的关键代码如下：

```
package entity.inherit.singletable;
…
@Entity
@DiscriminatorValue("STUDENT")
public class Student extends Person implements Serializable {
    private double scholarship;
    …
    public String toString(){
        return "Student: "+super.toString()+", scholarship="+scholarship;
    }
}
```

@DiscriminatorValue("STUDENT")指定当前的子类在关系数据库表中区分列的值为 STUDENT。

子类 Teacher 的关键代码如下：

```
package entity.inherit.singletable;
//…导包略
@Entity
@DiscriminatorValue("TEACHER")
public class Teacher extends Person implements Serializable {
    private double salary;
    //省略了 getter 和 setter 方法
```

```
    public String toString(){
        return "Teacher: "+super.toString()+", salary="+salary;
    }
}
```

@DiscriminatorValue("TEACHER ")指定当前的子类在关系数据库表中区分列的值为 TEACHER。

下面使用测试类 SingleTableTest 来完成实体 Student 和 Teacher 的持久化，其关键代码如下：

```
package entity.inherit.singletable;
…
public class SingleTableTest {
    public static void main(String[] args) {
        add();
        listAll();
    }
    public static void listAll() {
        EntityManagerFactory ef
            =Persistence.createEntityManagerFactory("firstJPA2.1");
        EntityManager em =ef.createEntityManager();
        String jpql ="from Person";
        Query query =em.createQuery(jpql);
        List<Person>all =query.getResultList();
        em.close();
        ef.close();
        for (Person p: all){
            System.out.println(p);
        }
    }

    public static void add() {
        EntityManagerFactory ef
            =Persistence.createEntityManagerFactory("firstJPA2.1");
        EntityManager em =ef.createEntityManager();
        em.getTransaction().begin();
        Student s =new Student();
        s.setName("Alice");
        s.setSex("woman");
        s.setAge(20);
        s.setScholarship(1000.00);
        em.persist(s);

        Teacher t =new Teacher();
```

```
        t.setName("John");
        t.setSex("man");
        t.setAge(40);
        t.setSalary(2000);
        em.persist(t);

        em.getTransaction().commit();
        em.close();
        ef.close();
    }
}
```

add()方法的作用是在数据库 person 表插入两条记录。listAll()方法是使用 JPA 查询所有 Person 实体，持久化提供者会查询 person 表的所有记录并在控制台打印结果如下：

```
Student: id=1, name=Alice, sex=woman, age=20, scholarship=1000.0
Teacher: id=2, name=John, sex=man, age=40, salary=2000.0
```

由此可见，所有的继承层次结构中的实体类对象都被输出了，展示了多态行为的结果。

整个类继承层次结构使用单个表的映射方式常用于继承结构中类的属性较少的情况。

该映射方式的优点有两个：

(1) 支持多态。

(2) 比其他继承映射策略的性能好。因为所有的实体属性都在一张表中，访问子类实体属性不用进行连接操作。

该映射方式的缺点有 3 个：

(1) 关系数据库必须提供单独的一列来区分继承层次结构中不同的类型。

(2) 子类属性对应的表的字段不能为 NOT NULL。

因为如果将某个子类字段定义为 NOT NULL，则在持久化那些不包含该属性的其他实体时会抛出异常。例如，如果将字段 salary 定义为 NOT NULL，那么在持久化 Student 对象时会抛出异常。

(3) 如果类的继承结构很深，那么表将含有大量字段。

10.5.2 各子类使用单独的表

每个实体类及其子实体类在关系数据库中都有对应表，每个对应表只含有各个类新增的属性对应的字段，不会包括其父类属性对应的字段。

该映射方式中 Person 类的关键代码如下：

```
package entity.inherit.separatetable;
```

```
...
@Entity
@Inheritance(strategy=InheritanceType.JOINED)
public class Person implements Serializable {
    @Id
    @GeneratedValue
    protected int id;
    private String name;
    private String sex;
    private int age;
...
    public String toString() {
        return "id="+id+", name="+name+", sex="+sex+", age="+age;
    }
}
```

在最上层的根类 Person 中使用@Inheritance(strategy=InheritanceType.JOINED)指定映射继承的方式为各子类使用单独表。

子类 Student 的关键代码如下：

```
package entity.inherit.separatetable;
...
@Entity
public class Student extends Person implements Serializable {
    private double scholarship;
    ...
    public String toString(){
        return "Student: "+super.toString()+", scholarship="+scholarship;
    }
}
```

子类 Teacher 的关键代码如下：

```
package entity.inherit.separatetable;
...
@Entity
public class Teacher extends Person implements Serializable {
    private double salary;
    //…getter 和 setter 方法略
    public String toString(){
        return "Teacher: "+super.toString()+", salary="+salary;
    }
}
```

在关系数据库中每个实体类 Person、Student、Teacher 都有对应表，其 DDL 如下：

```
CREATE TABLE `person` (
  `id` int(11) NOT NULL AUTO_INCREMENT,
  `age` int(11) NOT NULL,
  `name` varchar(255) DEFAULT NULL,
  `sex` varchar(255) DEFAULT NULL,
  PRIMARY KEY (`id`)
) ENGINE=InnoDB AUTO_INCREMENT=3 DEFAULT CHARSET=utf8;

CREATE TABLE `student` (
  `scholarship` double NOT NULL,
  `id` int(11) NOT NULL,
  PRIMARY KEY (`id`),
  CONSTRAINT `FK_ohs43dct8k52ch2exlmf4bs3l` FOREIGN KEY (`id`) REFERENCES
`person` (`id`)
) ENGINE=InnoDB DEFAULT CHARSET=utf8;

CREATE TABLE `teacher` (
  `salary` double NOT NULL,
  `id` int(11) NOT NULL,
  PRIMARY KEY (`id`),
  CONSTRAINT `FK_g6jmt7fcm6gfd0jvhimb9xy84` FOREIGN KEY (`id`) REFERENCES
`person` (`id`)
) ENGINE=InnoDB DEFAULT CHARSET=utf8;
```

各个表都有 id 字段；person 表只包含 Person 类属性对应的字段；student 表和 teacher 表只包括 Student 和 Teacher 实体类各自新增的属性对应的字段。在解析单个实体时需要使用该字段完成连接操作。

各子类使用单独的表的映射方式常用于继承结构中类的属性较多的情况。

该映射方式的优点有两个：

(1) 支持多态；

(2) 不用为任何表的任何字段定义 NOT NULL 约束。

该映射方式的缺点主要是必须通过连接表才能获得子类的所有属性。因此，如果类继承层次结构很深，访问子类实体属性需要更多的连接操作，会严重影响系统的性能。

10.5.3 各个具体实体类使用单个表

每个实体类在数据库中都有对应的一张表；每张表中包含了对应实体类的新增属性及从父类继承来的所有属性。

Person 类的关键代码如下：

```
package entity.inherit.singletableperconcrete;
...
@Entity
```

```
@Inheritance(strategy =InheritanceType.TABLE_PER_CLASS)
public class Person implements Serializable {
    @Id
    @GeneratedValue(strategy =GenerationType.TABLE)
    protected int id;
    private String name;
    private String sex;
    private int age;
    ...
    public String toString() {
        return "id=" +id +", name=" +name +", sex=" +sex +", age=" +age;
    }
}
```

在最上层的根类中使用@Inheritance(strategy＝InheritanceType. TABLE_PER_CLASS)指定映射继承的方式为各个具体实体类使用单个表。对于 MySQL 5 数据库，Person 实体类的主键生成策略使用的是 GenerationType. TABLE，不能像以前那样使用自动主键生成方式。

子类 Student 的关键代码如下：

```
package entity.inherit.singletableperconcrete;
//…导包略
@Entity
public class Student extends Person implements Serializable {
    private double scholarship;
    // getter()和 setter()方法略
    public String toString(){
        return "Student: "+super.toString()+", scholarship="+scholarship;
    }
}
```

子类 Teacher 的关键代码如下：

```
package entity.inherit.singletableperconcrete;
...
@Entity
public class Teacher extends Person implements Serializable {
    private double salary;
    //getter()和 setter()方法略
    public String toString(){
        return "Teacher: "+super.toString()+", salary="+salary;
    }
}
```

以上各子类使用单独表的映射方式，在关系数据库中，每个实体类 Person、Student、Teacher 都有对应的表，其 DDL 如下：

```
CREATE TABLE `person` (
  `id` int(11) NOT NULL,
  `age` int(11) NOT NULL,
  `name` varchar(255) DEFAULT NULL,
  `sex` varchar(255) DEFAULT NULL,
  PRIMARY KEY (`id`)
) ENGINE=InnoDB DEFAULT CHARSET=utf8;

CREATE TABLE `student` (
  `id` int(11) NOT NULL,
  `age` int(11) NOT NULL,
  `name` varchar(255) DEFAULT NULL,
  `sex` varchar(255) DEFAULT NULL,
  `scholarship` double NOT NULL,
  PRIMARY KEY (`id`)
) ENGINE=InnoDB DEFAULT CHARSET=utf8;

CREATE TABLE `teacher` (
  `salary` double NOT NULL,
  `id` int(11) NOT NULL,
  `age` int(11) NOT NULL,
  `name` varchar(255) DEFAULT NULL,
  `sex` varchar(255) DEFAULT NULL,
  PRIMARY KEY (`id`),
  CONSTRAINT `FK_g6jmt7fcm6gfd0jvhimb9xy84` FOREIGN KEY (`id`) REFERENCES
`person` (`id`)
) ENGINE=InnoDB DEFAULT CHARSET=utf8;

CREATE TABLE `hibernate_sequences` (
  `sequence_name` varchar(255) DEFAULT NULL,
  `sequence_next_hi_value` int(11) DEFAULT NULL
) ENGINE=InnoDB DEFAULT CHARSET=utf8;
```

student 表和 teacher 表包括 Student 和 Teacher 实体类各自新增的属性对应的字段和从父类继承的所有属性对应的字段。

hibernate_sequences 表是主键生成策略 GenerationType. TABLE 后用于保存主键的表。

该映射方式的优点有两个：

(1) 与类整个继承层次结构使用单个表相比，该继承映射策略允许在子类属性上定义 NOT NULL 约束；

(2) 使用该映射策略可以实现遗留数据库的映射。

该映射方式的缺点主要是不支持多态，因此在查询子类属性时需要使用多个查询才能加载一个多态的实体关系。

10.5.4 实体继承总结

实体继承存在如下 4 种情况。

1. 实体类继承非实体类

一个实体类可以继承一个非实体类,且该非实体父类不是抽象类。

1) 在父类上使用注解@MappedSuperclass

在父类上使用注解@MappedSuperclass,JPA 不会为父类创建对应的数据库表;非实体超类不能作为参数传递给 EntityManager 或 Query 接口进行操作;子类继承的父类的属性可以持久化到子类对应的数据库表中;对子类查询时可以查询到这些继承的属性。

2) 在父类上不使用注解@MappedSuperclass

非实体父类前不使用注解@MappedSuperclass,JPA 不会为父类创建对应的数据库表;非实体超类不能作为参数传递给 EntityManager 或 Query 接口进行操作;子类继承的父类的属性不会被持久化;对子类查询时不会查询到这些继承的属性。

2. 非实体类继承实体类

JPA 允许非实体类可以继承实体类。JPA 仅仅将是父类进行映射,不会对非实体的子类进行映射。

3. 实体类继承抽象实体类

抽象类可以使用@Entity 注解或 XML 描述为一个实体。抽象实体和具体实体的唯一区别是抽象实体不能够被直接实例化。抽象实体能够被映射到数据库并能够作为查询对象。

4. 实体类继承其他实体类

这种情况在 10.5.1～10.5.3 节进行了详细说明。

10.6 JPA 查询语言

JPA 查询语言(Java Persistence API Query Language)简称 JPA QL,是一种与数据库无关的、基于实体的查询语言。JPA QL 支持投影、批量更新和批量删除、子查询、连接、分组、HAVING 操作。所有的 JPA QL 操作都在命名查询和动态查询中得到支持。另外,JPA QL 支持多态、支持类似 SQL 的函数功能,如 MAX()和 MIN()。

10.6.1 查询单个实体

查询单个实体表示查询实体类对应的数据库表的全部记录。

1. 定义查询语句

查询单个实体语句的一般格式如下：

```
SELECT o FROM Entity AS o
```

该语句的含义为查询实体 Entity 的所有对象。其中，SELECT、FROM、AS 是 JPA QL 的关键字。FROM 后面是查询的实体名，通常是实体类的类名。AS o 是将 o 作为实体 Entity 的别名。与 SQL 类似，AS 可以省略。

例如，"String jpql = "SELECT u FROM User u";"表示查询 User 实体即查询对应的 user 表的全部记录。

查询语句是 String 类型。查询语句中的 User 是实体类的类名，也可以使用类的限定名(即包名.类名)如：String jpql = "SELECT u FROM entity.User u"。

2. 创建 Query 对象

EntityManager 接口定义了创建 Query 对象的方法。

```
Query query =em.createQuery(jpql);
```

3. 执行查询

```
List<User>  allUsers =query.getResultList( );
```

以上是对 User 实体的查询，将返回包含所有的 User 实体对象的 List，List 中的每个 User 实体对象对应一条数据库 user 表的记录。如果查询结果为空，那么返回的 List 是空的。

还可以将以上 3 个步骤用一行语句表示：

```
List<User>  allUsers =em.createQuery("SELECT u FROM User u").getResultList( );
```

如果查询的结果只有一个实体，那么可以使用 EntityManager 的 getSingleResult() 方法：

```
User user = (User) em.createQuery("SELECT u FROM User u where id =1")
    .getSingleResult();
```

10.6.2 查询实体属性和关系属性(投影)

JPA QL 允许 SELECT 子句返回任意数量的基本属性或关系属性。

1. 查询实体的一个属性

查询实体的某个属性需要在 SELECT 子句中指定属性的名字，查询的返回值类型取决于被访问字段的类型。

例如，查询 User 实体的 name 属性的查询语句如下：

```
String jpql ="SELECT u.name FROM User AS u";
List<String>names =em.createQuery(jpql).getResultList();
```

name 属性是 String 类型，因此通过 Query. getResultList()方法得到满足查询条件的字符串 List＜String＞ names。

2. 查询实体多个属性

SELECT 子句查询多个属性时，getResultList()方法返回 java. util. List 类型的对象数组(Object []对象)。

例如，查询 User 实体的多个属性如下：

```
String jpql ="SELECT u.name, u.id FROM User AS u";
final List<Object[]>properties =em.createQuery(jpql).getResultList();
```

3. 访问关联关系中的实体属性

访问关联关系中的实体属性需要使用.运算符。

例如，Employee 与 Department 是多对一关系，其实体关系如下所示：

```
@Entity
public class Employee{
    @ManyToOne
    private Departmentdepartment;
...
}
```

查询员工所在部门的查询语句如下：

```
SELECT e.department.name FROM Employee AS e
```

10.6.3 使用 IN 访问关系集合属性

在关系映射中直接获取关系集合中的元素是非法的。例如，Employee 与 Phone 是双向一对多关系，以下获取 Employ 的所有电话的 JPA QL 查询语句是非法的：

```
SELECT e.phones.number FROM Employee AS e;
```

获取关系集合元素需要在 FROM 子句中使用 IN 操作符指定一个唯一标识符，并在查询语句中选择该标识符。使用 IN 操作符获取 Employee 的所有电话的查询语句如下：

```
String jpql ="SELECT p FROM Employee AS e, IN( e.phones ) p";
```

以下代码使用 IN 操作符获取 Employee 的所有电话。

```
package test;
…
public class UserTest {
    public static void main(String[] args) {
        query ();
    }
    …
    @SuppressWarnings("unchecked")
    public static void query() {
        EntityManagerFactory emfactory =
            Persistence.createEntityManagerFactory("firstJPA2.1");
        EntityManager em =emfactory.createEntityManager();
        String jpql ="SELECT p FROM Employee AS e, IN( e.phones ) p";
        List<Phone>phones =em.createQuery(jpql).getResultList();
        for (Phone p : phones) {
            System.out.println(p.getNumber());
        }
        em.close();
        emfactory.close();
    }
    …
}
```

还可以使用.运算符直接获取集合关系属性的元素。以下语句使用.运算符直接获取 Employee 的集合关系属性 phones 的元素 number。

```
SELECT p.number FROM Employee AS e, IN( e.phones ) p;
List<String>  phones =em.createQuery(jpql).getResultList();
```

还可以使用连接操作 INNER JOIN 查询雇员的所有电话号码，关键代码如下：

```
@SuppressWarnings("unchecked")
public static void query2() {
    EntityManagerFactory emfactory =
        Persistence.createEntityManagerFactory("firstJPA2.1");
    EntityManager em =emfactory.createEntityManager();
    String jpql ="SELECT p.number FROM Employee e INNER JOIN e.phones p";
    List<String>phones =em.createQuery(jpql).getResultList();
    for (String number : phones) {
        System.out.println(number);
    }
```

```
        em.close();
        emfactory.close();
    }
```

10.6.4 连接实体

JPA QL 支持类似 SQL 中的连接(JOIN)操作。

1. INNER JOIN

JPA QL 默认采用内连接(INNER JOIN)完成记录获取。内连接用于返回在关系实体中存在匹配值的所有实体。它要求 INNER JOIN 右边的表达式必须返回实体。

例如,以下语句查询至少存在一个雇员的所有部门集合。

```
SELECT d FROM Department d JOIN d.employees e
```

以下语句与上面的语句有等价的效果:

```
SELECT d FROM Department d INNER JOIN d.employees e
```

2. LEFT JOIN

左连接(LEFT JOIN 或 LEFT OUT JOIN)用于返回左边实体集合中的所有实体,即使右边的实体集合不存在匹配的实体。

以下语句使用 LEFT JOIN 查询所有的部门,即使部门没有一个员工。

```
SELECT d FROM Department d LEFT JOIN d.employees e
```

3. FETCH JOIN

JOIN FETCH 语法允许预先加载返回实体的关系,即使关系属性设置为 FetchType.LAZY。

1) N+1 问题

通常在一对多关系映射中,多方采用懒加载方式,这会导致存在 N+1 问题。例如,雇员和电话是一个一对多关系,其关系如下所示:

```
@OneToMany (fetch=FetchType.LAZY)
private Collection<Phone>phones;
```

如果需要打印出所有雇员信息(包括其电话号码),可以采用查询所有的雇员再通过遍历雇员列表的方式进行:

```
Query query =manager.createQuery("SELECT e FROM Employee e");
List results =query.getResultList();
Iterator it =results.iterator();
```

```
while (it.hasNext( )) {
    Employee e = (Employee)it.next();
    System.out.print(e.getName());
    for (Phone p : e.getPhones()) {
        System.out.print(p.getNumber());
    }
    System.out.println("");
}
```

以上的这种方式存在如下问题：

- 性能问题。在 Employee 类中 Phone 被声明为延迟加载，所以在进行初始查询时 Phone 集合不会同时进行查询得到结果。
- 当执行 getPhones()方法时，持久化引擎必须做额外的工作以获取与 Employee 关联的 Phone 实体。这被称为 N+1 问题。因为必须进行超过最初查询的 N 次额外查询。在基于数据库的应用中，应当尽可能减少应用与数据库的交互次数。

2）使用 JOIN FETCH 语法预加载多方实体对象改进 N+1 问题

以下语句使用 JOIN FETCH 语法预加载多方 Phone 对象改进 N+1 问题。

```
SELECT e FROM Employee e LEFT JOIN FETCH e.phones
```

LEFT JOIN FETCH 会预先加载关联的 Phone 对象。与 N+1 问题查询比较，此处只有一次数据库查询，会极大地提高效率。

10.6.5 使用参数

与 JDBC 中的 java.sql.PreparedStatement 类似，JPA QL 允许在查询语句中使用参数，以便在使用不同的参数执行多次执行查询时重用查询语句。JPA QL 查询语句中可使用命名参数和位置参数。

1. 使用命名参数

命名参数格式为：

```
:命名参数
```

例如，在用户查询时，使用用户名作为命名参数的查询步骤如下：

```
// 定义查询语句
String jpql ="select d from Department d where id=:departmentID";
// 创建 Query 对象
Query query =em.createQuery(jpql);
// 设置命名参数
query.setParameter("departmentId", new Integer(1));
// 执行查询获得结果
Department d= (Department) query.getSingleResult();
```

```
// 通过调用 size()方法加载关联对象
d.getEmployees().size();
```

方法 setParameter("departmentId ", new Integer(1))用于为 departmentId 参数赋值为大小为 1 的 Integer 对象。由于 Java SE 5 实现了基本类型与基本类型包装类的自动相互转换(称为自动装箱和自动出箱),因此此处可以改为如下语句:

```
query.setParameter("departmentId",1);
```

2. 使用位置参数

位置参数格式:

```
?参数索引
```

使用位置参数实现用户查询的步骤如下:

```
// 定义查询语句
String jpql ="select d from Department d where id=?1";
// 创建 Query 对象
Query query =em.createQuery(jpql);
// 设置命名参数
query.setParameter(1, new Integer(1));
// 执行查询获得结果
Department d=(Department) query.getSingleResult();
// 通过调用 size()方法加载关联对象
d.getEmployees().size();
```

setParameter(1, new Integer(1))用于为第一个参数 departmentId 赋值为大小为 1 的 Integer 对象。

3. 设置日期参数

数据库中表示时间日期的数据类型通常有 DATE、TIME、TIMESTAMP。Java 中与数据库时间日期类型对应的是 java. sql. Date、java. sql. Time、java. sql. TimeStamp。Java 中除了访问数据库之外所使用的时间日期类型有 java. util. Date、java. util. Calendar。

如果需要在查询中传递 java. util. Date、java. util. Calendar 类型的参数访问数据库,需要使用特殊的 setParameter()方法,该方法定义在 Query 接口中。数据库中与时间日期有关的 Date、Time、Timestamp 封装在 TemporalType 枚举中。

```
package javax.persistence;
public enum TemporalType{
    DATE, //java.sql.Date
    TIME, //java.sql.Time
    TIMESTAMP //java.sql.Timestamp
}
```

```
public interface Query{
    Query setParameter(String name, java.util.Date value,
        TemporalType temporalType);
    Query setParameter(String name, Calendar value,
        TemporalType temporalType);
    Query setParameter(int position, Date value, TemporalType temporalType);
    Query setParameter(int position, Calendar value,
        TemporalType temporalType);
}
```

这些特殊的 setParameter()方法有 3 个参数：第一个是命名参数，第二个是在程序中使用的时间日期类型 java. util. Date 或 java. util. Calendar，第三个是第二个参数对应的 Java 数据库时间日期类型。

例如，数据库中某个字段是 DATE 类型，Java 程序中使用的是 java. util. Date 类型，使用 JPA 设置日期参数为 java. util. Date 类型。

```
//定义查询语句
String jpql ="select d from Department d where birthday=:birthday";
//创建 Query 对象
Query query =em.createQuery(jpql);
//设置命名参数为 java.sql.Date
query.setParameter(birthday, new java.util.Date(),TemporalType.Date);
…
```

如果需要设置日期参数为 java. util. Calendar 类型，那么可将上述最后一句代码修改如下：

```
query.setParameter(birthday, new java.util. Calendar (),TemporalType.Date);
```

10.6.6 分页功能

Query API 提供了 setMaxResults()和 setFirstResult()来实现分页功能。setMaxResults(max)方法设置从数据库返回的查询结果的最大记录数为 max，即每页显示多少条记录；setFirstResult(index)方法设置从数据库返回的查询结果的第一条记录的索引是 index(索引从 0 开始)。

使用分页功能获取 Employ 表的记录的关键代码如下：

```
public List getEmployees(int max, int index) {
    Query query =entityManager.createQuery("from Employee e");
    return query.setMaxResults(max).setFirstResult(index).getResultList();
}
```

10.6.7 ORDER BY

ORDER BY 子句用于将查询后结果集中实体进行排序。ORDER BY 子句的含义同 SQL 中的一致。

例如，对 Employee 实体的查询中使用 ORDER BY 的查询语句如下：

```
SELECT e FROM Employee AS e ORDER BY e.name
```

这将会返回一个名字按照字母排序的雇员实体集合。

ORDER BY 对结果集排序有升序(ASC)和降序(DESC)，默认是升序，即排序后的结果集中每个项目按照从小到大的顺序排列。

以下对 Employee 实体查询语句中使用 ORDER BY 中显式地指示排序类型是升序：

```
SELECT e FROM Employee AS e ORDER BY e.name ASC
```

以下对 Employee 实体查询语句中使用 ORDER BY 中显式地指示排序类型是降序：

```
SELECT e FROM Employee AS e ORDER BY e.name DESC
```

10.6.8 DISTINCT

DISTINCT 关键字用于排除查询结果中重复的记录。

例如，下面的查询可能会返回重复的记录：

```
SELECT c FROM Student s INNER JOIN s.courses c
```

如果课程属于多个学生，那么查询结果中会出现重复的课程。使用 DISTINCT 关键字能排除返回结果集中的重复记录。

```
SELECT DISTINCT c FROM Student s INNER JOIN s.courses c
```

如下代码使用 DISTINCT 关键字排除查询结果中重复的记录：

```
public List<Course>listAllCourses(){
    String jpql ="select DISTINCT c from Student s inner join s.courses c";
    Query query =em.createQuery(jpql);
    List<Course>allCourses =query.getResultList();
    return allCourses;
}
```

10.6.9 在查询中构建对象

JPA 支持在 SELECT 语句中将查询结果传递给实体构造方法以创建一个新的对象。

例如，如下代码在 SELECT 语句中构造 User 对象：

```
String jpql ="SELECT new User(u.id,u.name) FROM Useru";
Query query =em.createQuery(jpql);
List<User>result =(List<User>)query.getResultList();
```

Query 对象会根据每次返回值将 id 和 name 传递给 User 的构造方法，自动创建一个 User 对象。执行查询语句后返回了一个 User 对象的列表。

10.6.10 批量更新和批量删除

JPA QL 支持批量更新和批量删除。批量更新是一次性更新多条记录；批量删除是一次性删除多条记录。

批量更新和批量删除操作遵循如下规则：

(1) 操作作用于指定的实体及其所有的子类；

(2) 操作不会作用于级联到关联的实体；

(3) 批量更新操作中更新的实体的属性类型必须与数据库中的字段类型匹配；

(4) 批量操作直接在数据库中进行。因此，批量操作脱离了乐观锁的监控，并且存在的 version 列也不会被自动更新；

(5) 持久化上下文不会与批量操作结果进行同步。

在进行批量更新和删除时可能产生数据库和持久化上下文中实体状态的不一致。因此，推荐执行批量更新和删除操作在事务一开始进行或在单独的事务中进行。另外，在批量操作之前执行 EntityManager. flush()方法和 EntityManager. clear()方法可以确保更加安全。

以下语句中使用批量更新操作设置用户名全部都是 Mike。

```
UPDATE User u SET u.name = 'Mike'
```

该批量更新语句将 User 实体对应的数据库表 user 的所有 name 列更新为 Mike。

10.6.11 使用 WHERE 子句

与 SQL 中的 WHERE 子句一样，JPA QL 中的 WHERE 子句用于条件查询。

如下语句使用 WHERE 子句查找名字是 Alice 的所有员工：

```
SELECT e FROM Employee AS e WHERE e.name = 'Alice'
```

10.6.12 GROUP BY 和 HAVING

GROUP BY 将查询结果按照某个属性或者多个属性的值进行分组，值相等的为一组。

以下查询语句中使用 GROUP BY 进行分组查询，获得女员工的数目。

```
SELECT e.sex, COUNT(e) FROM Employee e GROUP BY e.sex HAVING e.sex = 'F'
```

该语句首先使用 GROUP BY 子句按照 e. sex 进行分组，再用聚集函数 COUNT 对每一组计数。HAVING 短语指定了选择组的条件，只有员工性别是 F(女)的组才被选出。

10.6.13 NativeQuery

JPA QL 提供了 NativeQuery 接口执行原生 SQL 语句。

以下代码用于执行原生 SQL 语句,更新数据库 user 表中 id 为 1 的记录的 name 字段值为 John。

```
public void update(){
    //下面语句是原生 SQL 语句
    String sql ="update employee set name ='John ' where id=1";
    Query query =em.createNativeQuery(sql);
    query.executeUpdate();
    …
}
```

如果需要使用原生 SQL 查询表并将结果映射到实体,EntityManager 提供了如下方法:

```
createNativeQuery(String sql, Class resultClass)
```

以下代码使用原生 SQL 语句查询 user 表的所有记录,并将结果映射到实体。

```
public List<Employee>listAllEmployees(){
    String sql ="select * from user";//此句是原生 SQL 语句
    Query query =em.createNativeQuery(sql,Student.class);
    //可以将结果映射到实体
    List<Student>all =query.getResultList();
    return all;
}
```

10.6.14 命名查询

命名查询又叫静态查询。在进行查询之前在实体中使用@NamedQuery 注解定义一个或多个命名查询;进行查询时使用 EntityManager 创建 NamedQuery 对象进行查询。

以下代码使用命名查询获取所有的 Employee 实体对象。

首先在实体 Employee 中定义了命名查询语句如下:

```
@Entity
@NamedQuery(name="listAll" queryString="SELECT e FROM Employee e")
public class Employee implements Serializable{…}
```

@NamedQuery 注解的 name 属性指定命名查询的名字是 listAll,queryString 属性指定命名查询语句,该查询语句返回所有的 Employee 实体对象。

如果需要定义多个命名查询，则需要使用@NamedQueries 注解，并在该注解中使用多个@NamedQuery 注解。

以下代码定义两个命名查询：一个命名查询用于查询全部 Employee，另一个通过 id 获取一个 Employee。

```
@Entity
@NamedQueries({
    @NamedQuery(name="listAll" queryString="SELECT a FROM Employee e"),
    @NamedQuery(name="getEmployeeById"
        queryString="SELECT a FROM Employee e WHERE id=:id")
})
public class Employee implements Serializable{…}
```

以下代码是使用 EntityManager 创建 NamedQuery 对象执行查询。

```
public List<Employee>listAllEmployees(){
    Query query =em.createNamedQuery("listAll");
    List<Student>all =query.getResultList();
    return all;
}
```

listAllEmployees 方法中首先通过 createNamedQuery("listAll")创建 NamedQuery 对象，接着后通过 getResultList()获取查询结果。

10.6.15 调用存储过程

JPA QL 调用存储过程的语句格式：

```
{CALL 存储过程名(参数 1,参数 2,…)}
```

通过 EntityManager 的 createNativeQuery()方法创建 NativeQuery 对象并调用 NativeQuery 的 executeUpdate()方法调用执行存储过程。

在 JPA QL 中可以调用两种形式的存储过程。

1. 调用无返回值的存储过程

无返回值的存储过程 AddEmployee 的功能是向数据库插入一条记录，其 DDL 如下：

```
CREATE PROCEDURE 'AddEmployee'()
  NOT DETERMINISTIC
  SQL SECURITY DEFINER
  COMMIT
BEGIN
  INSERT INTO employee('name', 'sex') values('John', 'M');
END;
```

以下代码是 JPA 调用无返回值的存储过程：

```
public void addEmployee(){
    Query query =em.createNativeQuery("{CALL AddEmployee()}");
    query.executeUpdate();
}
```

2. 调用返回单值的存储过程

返回单值的存储过程 getEmployeeName 的 DDL 如下：

```
CREATE PROCEDURE 'getEmployeeName'(IN id INTEGER(11))
  NOT DETERMINISTIC
  SQL SECURITY DEFINER
    COMMIT
BEGIN
    SELECT name FROM employee WHERE 'id'=id;
END;
```

以下代码是 JPA 调用返回单值的存储过程：

```
public String getEmployeeName (){
    Query query =em.createNativeQuery("{CALL getEmployeeName (?)}");
    query.setParameter(1, new Integer(1));
    String result =query.getSingleResult().toString();
    StringBuffer out =new StringBuffer("----打印结果----<BR>");
    out.append("员工名字为:"+result+"<BR>");
    return out.toString();
}
```

以上代码分别创建 NativeQuery 对象；使用 setParameter(1, new Integer(1))设置参数；调用 getSingleResult(). toString()方法执行存储过程后得到查询结果并转型为 String；使用 StringBuffer 类型的 out 对象保存查询的单值结果，最后返回一个 out 对象的字符串。

小　结

本章首先介绍了与 O/R 映射、JPA 等相关概念。接着介绍了实体的概念和实体类的编写规范，使用 EntityManager 接口相关方法对实体进行增、删、改、查等操作，使用 JPA 实现关系映射和映射继承。最后介绍的是使用 JPA 查询语言实现实体查询、实体单个和多个属性查询、条件查询、分组查询、消除重复记录等。

思考与习题

1. Java 应用将数据持久化到关系数据库可以使用哪些方式?

2. 什么是 O/R 映射? 什么是 JPA? 这二者有何区别?

3. 简要描述在 Eclipse 下搭建 JPA Java SE 环境的步骤。

4. 什么是实体? 实体类的编写规范有哪些?

5. 编写一个实体类 Student,该类有 int id(主键)、String code(学号)、String name(姓名)。要求尽可能显式使用注解及其属性完成,并使用注释说明各个注解的作用。

6. Java SE 环境中使用 JPA 完成对实体 Student 的增、删、改、查。

7. EntityManager 对象的管理方式有几种? 分别是什么?

8. 实体的状态有几种? 举例说明导致实体状态变迁的 EntityManager API。

9. 举例说明 EntityManager 的 refresh(Object obj)、flush()、setFlushMode(FlushModeType flushMode)方法的作用分别是什么。

10. JPA 为实体定义的生命周期事件有哪些? 在调用 EntityManager 的何种方法时会触发何种事件?

11. 举例说明分别使用实体类中定义实体生命周期回调方法和使用监听器类定义生命周期回调方法。

12. 关系数据库中的关系和 JPA 中的关系分别是哪些?

13. 举例说明单向一对一映射与双向一对一映射的区别。

14. 举例说明单向一对多映射与双向一对多映射的区别。

15. 举例说明单向多对多映射与双向多对多映射的区别。

16. JPA 映射继承的方式有几种? 分别是什么?

17. 使用 JPA QL 查询第 5 题中 Student 实体的 id 属性。

18. 使用 JPA QL 查询第 5 题中 Student 实体的 name 和 code 属性。

第11章

Spring 框架核心基础

Spring 是一个为方便和快速开发 Java 应用提供了全面的基础设施的开源 Java 框架。Spring 框架最初是由 Rod Johnson 编写，并且 2003 年 6 月首次在 Apache 2.0 许可下发布。本书介绍的 Spring 框架的知识基于 Spring 框架 4.1.6 版本。

11.1 Spring 框架简介

Spring 框架为企业级应用开发提供了一个轻量级解决方案，主要特点如下：

- 以依赖注入为核心；
- 支持 AOP 声明式事务管理；
- 支持与多种持久层技术整合；
- 支持与多种 Web MVC 框架整合；
- 提供了 Template 功能方便快速开发。

11.1.1 Spring 体系结构

Spring 框架提供了约 20 个模块，可以根据应用程序的要求来使用其中的部分模块。Spring 框架的体系结构如图 11-1 所示。

1. 核心容器

核心容器由 Beans、Core、Context、SpEL 模块组成。Core 模块提供了框架的基础，包括 IoC 和依赖注入功能。

Beans 模块提供的 BeanFactory 是一个工厂模式的经典实现。

Context 模块建立在 Core 和 Bean 模块基础上，可以访问定义的任何对象。ApplicationContext 接口是 Context 模块的重点。

SpEL 模块即 Spring 表达式语言模块，提供了在运行时查询和操作对象图的能力。

2. 数据访问/集成

数据访问/集成层包括 JDBC、ORM、OXM、JMS 和事务处理模块。

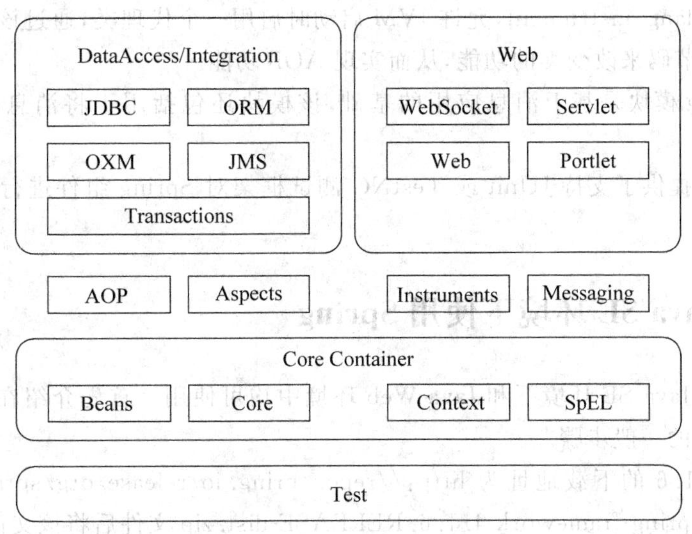

图 11-1 Spring 体系结构

JDBC 模块提供没有冗余的 JDBC 编码的 JDBC 抽象层。

ORM 模块为流行的对象关系映射框架(如 JPA、JDO、Hibernate、iBatis 等)提供了集成能力。

OXM 模块提供的抽象层支持对 JAXB、Castor、XMLBeans、JiBX、XStream 的对象/XML 映射实现。

JMS 模块提供对 Java 消息处理。

Transactions 模块提供面向 POJO 的编程式和声明式事务管理。

3. Web

Web 层由 Web、Web-MVC、Web-Socket、Web-Portlet 模块组成。

Web 模块提供了 Web 应用的基础功能,例如,多文件上传功能、使用 Servlet 监听器和 Web 应用的上下文来初始化 IoC 容器。

Web-MVC 模块提供了 Spring 自己的 MVC 实现。

Web-Socket 模块为 WebSocket-based 提供了支持,而且在 Web 应用程序中提供了客户端和服务器端之间通信的两种方式。

Web-Portlet 模块提供了基于 Web-MVC 的 Portlet 环境中的 MVC 实现。

4. 其他

其他一些重要的模块包括 AOP、Aspects、Instrumentation、Web 和测试模块。

AOP 模块提供了符合 AOP 联盟规范的面向方面的编程实现,允许定义方法拦截器和切入点对功能实现代码进行解耦合。

Aspects 模块提供了与功能强大且成熟的 AOP 框架 AspectJ 的集成。

Instrumentation 模块提供了对类 Instrumentation 的支持和类加载器的实现。Java

5.0 引入 java.lang.instrument，允许 JVM 启动时启用一个代理类，通过该代理类在运行时修改类的字节码来改变类的功能，从而实现 AOP 功能。

Messaging 模块是基于消息应用的基础，该模块还包括用于将消息映射到方法的注解。

Test 模块提供了支持 JUnit 或 TestNG 测试框架对 Spring 组件进行单元测试或集成测试。

11.1.2 Java SE 环境下使用 Spring

Spring 在 Java SE 环境下和 Java Web 环境中均可使用。首先介绍在 Java SE 环境下使用 Spring 的一般步骤。

Spring 4.1.6 的下载地址为 http://repo.spring.io/release/org/springframework/spring。下载 spring-framework-4.1.6.RELEASE-dist.zip 文件后将该文件解压。

Spring 默认使用 Apache Commons Logging 日志组件，其下载地址为 http://commons.apache.org/logging/。本书使用的是 commons-logging-1.2，下载后的文件是 commons-logging-1.2-bin.zip，将该文件解压。

1. 创建 Eclipse Java 项目

在 Eclipse 中创建 Java 项目 ch11-1。

2. 配置构建路径

配置项目 ch11-1 构建路径，添加 Spring 类库所有 jar 文件和 commons-logging-1.2.jar 文件。

3. 编写 Java 类并在 beans.xml 中配置

(1) 在项目 ch11-1 的 src 下新建类 entity.User，其关键代码如下：

```
package entity;
//…导包略
public class User {
    private int id;
    private String name;
    private String password;
    private int age;
    private Date birthday;
    …getter()和 setter()方法略
}
```

(2) 在项目的 src 下新建 beans.xml，使用 XML 配置方式配置 Bean user，其代码如下：

```
<?xml version="1.0" encoding="UTF-8"?>
<beans xmlns="http://www.springframework.org/schema/beans"
    xmlns:xsi="http://www.w3.org/2001/XMLSchema-instance"
    xsi:schemaLocation="http://www.springframework.org/schema/beans
        http://www.springframework.org/schema/beans/spring-beans.xsd">
    <bean id="user" class="entity.User">
        <property name="id" value="1"/>
        <property name="name" value="Mike"/>
        <property name="password" value="123"/>
        <property name="age" value="23"/>
    </bean>
</beans>
```

4. 创建测试类

新建 SpringTest 类进行测试,其代码如下:

```
package entity;
import org.springframework.context.ApplicationContext;
import org.springframework.context.support.ClassPathXmlApplicationContext;
public class SpringTest {
    public static void main(String[] args) {
        ApplicationContext context =
            new ClassPathXmlApplicationContext("beans.xml");
        User user = (User) context.getBean("user");
        System.out.println(user.getId() +"-" +user.getName() +"-"
            +user.getPassword() +"-"
            +user.getAge());
    }
}
```

ClassPathXmlApplicationContext()用于创建应用上下文、加载配置文件 beans. xml,并创建和初始化 beans. xml 配置文件中所有的对象。

context. getBean("user") 方法用于获得所需的 bean 对象 user,user 是在 beans. xml 的<bean id="user" class="entity. User">中定义的。

最后程序打印 user 对象的属性值。

5. 运行测试类

执行测试类 SpringTest 后 Eclipse 控制台输出结果如下:

```
1-Mike-123-23
```

11.2 IOC 容器

IOC 容器是 Spring 框架的核心。IOC 容器负责创建对象、管理对象生命周期，这些对象被称为 Spring Beans。IOC 容器使用依赖注入来管理组成一个应用程序的 Bean。

配置元数据描述了 IOC 容器需要实例化的对象信息。配置元数据可以使用 XML 文件配置或者 Java(代码)配置。ch11-1 项目中的 beans. xml 是一个使用 XML 配置元数据的例子。

Spring 提供了两种不同类型的 IOC 容器(以下简称容器)。

11.2.1 BeanFactory 容器

这是一个最简单的容器，它的主要功能是为依赖注入提供支持。这个容器由 org. springframework. beans. factory. BeanFactor 定义。在 Spring 中有大量 BeanFactory 接口的实现，最常使用的是 XmlBeanFactory 类。XmlBeanFactory 容器从一个 XML 文件中读取配置元数据。BeanFactory 用于轻量级应用，比如移动设备或者基于 applet 的应用程序。

11.2.2 ApplicationContext 容器

ApplicationContext 是 Spring 中较高级的容器。它除了负责 Bean 的创建和生命周期管理，还具有企业应用所需要的功能，例如，从属性文件解析文本信息和将事件传递给指定的监听器等。这个容器由 org. springframework. context. ApplicationContext 接口定义，包含 BeanFactory 所有的功能。一般情况下会优先使用 ApplicationContext 容器。

经常使用的 ApplicationContext 接口的实现主要如下：

(1) FileSystemXmlApplicationContext：该容器从 XML 文件中加载已被定义的 Bean。需要为其构造方法提供 XML 文件所在文件系统中的完整路径。

(2) ClassPathXmlApplicationContext：该容器从 XML 文件中加载已被定义的 Bean。需要为其构造方法提供位于 CLASSPATH 下的 XML 文件路径。

(3) WebXmlApplicationContext：该容器会在一个 Web 应用的范围内加载在 XML 文件中定义的 Bean。

在 ch11-1 项目中使用了 ClassPathXmlApplicationContext 容器。如下代码说明了如何使用 FileSystemXmlApplicationContext。

```
public class SpringTestByFileSystemXmlApplicationContext{
    public static void main(String[] args) {
        ApplicationContext context =new FileSystemXmlApplicationContext(
          "C:/win10-JavaProgram/eclipse-jee-mars-2-win32-x86_64/"
```

```
                +"apps/ch12-01-02/src/beans.xml");
        User user = (User) context.getBean("user");
        System.out.println(user.getId() +"-"
            +user.getName() +"-"
            +user.getPassword() +"-"
            +user.getAge());
    }
}
```

11.3 依赖注入

当某个A类的对象a依赖另外一个B类的对象b的时候,传统的应用程序是在A类中使用关键字new创建B类的对象b。

Spring提出了依赖注入的思想,即依赖对象b不由程序员实例化,而是通过Spring容器创建指定实例并且将实例注入需要该实例的对象中。

依赖注入的另一种说法是控制反转。通俗的理解是,平常new一个实例,这个实例的控制权是程序员;而控制反转是指new实例工作不由程序员来做而是交给Spring容器来做。

Spring依赖注入方式包括setter注入、构造方法注入、工厂方法注入。Spring中常用的是前两种。

本节使用XML配置来介绍setter注入和构造方法注入。

11.3.1 setter注入

setter注入是指使用属性的setter()方法注入Bean的属性值或依赖对象。setter注入灵活性高,是Spring实际应用中采用较多的注入方式。

setter注入要求Bean有一个无参数的构造方法和注入属性的setter方法。Spring通过Bean无参数的构造方法实例化Bean对象后,再通过反射机制调用setter()方法注入属性值。

Spring的XML配置文件beans.xml中,<beans>是根元素。<beans>中包含0个或多个<bean>元素,Spring使用<bean>元素配置一个Bean。

例如,user Bean所属的类entity.User具有id、name、password、age属性及其对应的getter和setter方法。在beans.xml中,user Bean的配置如下:

```
…
<beans …>
    <!--以下使用 bean 元素配置 bean -->
    <bean id="user" class="entity.User">
        <property name="id" value="1"/>
```

```
        <property name="name" value="Mike"/>
        <property name="password" value="123"/>
        <property name="age" value="23"/>
    </bean>
</beans>
```

<bean>元素的 id 属性指定 Bean 的唯一标识，Spring 通过 id 属性值访问该 Bean 实例。

<bean>元素的 class 属性指定 Bean 的实现类或接口。此处是 user Bean 的实现类 entity. User。

<bean>元素的子标签<property>的 name 和 value 属性分别指定 Bean 的属性名和属性值。

Spring 容器读取 beans. xml 文件，通过反射机制创建 User 的实例 user，并通过其 setter 方法为属性赋值。

11.3.2 构造方法注入

构造方法注入是指通过构造方法注入 Bean 的属性或依赖对象。构造方法注入是 Spring 另外一种常用的注入方式，需要使用<constructor-arg>标签配置构造方法需要的参数。构造方法注入分为 4 种方式。

1. 按类型匹配参数

例如，User 类有一个带两个参数的构造方法，其关键代码如下：

```
package entity1;
public class User {
    ...
    public User(int id, String name) {
        this.id =id;
        this.name =name;
    }
    ...
}
```

使用按类型匹配参数方式注入 User 的属性值的配置如下：

```
<bean id="user" class="entity1.User">
  <constructor-arg type="int">
    <value>1</value>
  </constructor-arg >
  <constructor-arg type="java.lang.String">
    <value>Mike</value>
  </constructor-arg >
```

```
</bean>
```

<constructor-arg type="int">表示配置 User 类构造方法的一个 int 类型的参数；<value>1</value>表示该参数的值为 1。

<constructor-arg type="java. lang. String">表示配置 User 类构造方法的一个 String 类型的参数；<value>Mike</value>表示参数值为 Mike。

注意：多个<constructor-arg>的书写顺序并不表示构造方法中的参数顺序。因此，对于以上的 User 类，如下配置也是正确的。

```
<bean id="user" class="entity1.User">
   <constructor-arg type="java.lang.String">
      <value>Mike</value>
   </constructor-arg >
   <constructor-arg type="int">
      <value>1</value>
   </constructor-arg >
</bean>
```

2. 按索引匹配参数

例如，User 类有一个带两个参数的构造方法，其关键的代码如下：

```
package entity2;
public class User {
   …
   public User(String name, String password) {
      this.name =name;
      this.password =password;
   }
   …
}
```

由于这两个参数都是 String 类型，所以使用如下按照类型匹配参数的配置方式就不能确定对应关系。

```
<bean id="user"class="entity2.User">
   <constructor-arg type=" java.lang.String">
      <value>Mike</value>
   </constructor-arg >
   <constructor-arg type="java.lang.String">
      <value>123</value>
   </constructor-arg >
</bean>
```

这时可以使用<constructor-arg>的 index 指定对应顺序。正确的配置如下：

```
<bean id="user" class="entity2.User">
   <constructor-arg type="java.lang.String" index="0">
      <value>Mike</value>
   </constructor-arg>
   <constructor-arg type="java.lang.String" index="1">
      <value>1</value>
   </constructor-arg>
</bean>
```

3. 同时使用类型和索引匹配参数

有些情况下需要同时使用 type 和 index 来匹配参数。

例如,User 类有两个构造方法,其关键代码如下:

```
package entity3;
public class User {
    ...
    public User(int id, String name, String password) {
        this.id =id;
        this.name =name;
        this.password =password;
    }
    public User(int id, String name, int age) {
        this.id =id;
        this.name =name;
        this.age =age;
    }
    ...
}
```

如果仅仅使用 index 就不能满足要求。对于 User 类第二个构造方法,正确配置如下:

```
<bean id="user" class="entity3.User">
   <constructor-arg index="0" type="int">
      <value>1</value>
   </constructor-arg>
   <constructor-arg index="1" type="java.lang.String">
      <value>Mike</value>
   </constructor-arg>
   <constructor-arg index="2" type="int">
      <value>23</value>
   </constructor-arg>
</bean>
```

4. 通过自身类型反射匹配参数

构造方法参数不是基本数据类型且各个参数都是不同类型，由于 Java 的反射机制可以获取构造方法参数的类型，这种情况下即使不配置 type 和 index，Spring 依然可以正确完成构造方法参数注入。

例如，如下的 User 类有带两个参数的构造方法，参数分别是 Integer 和 String。

```
package entity4;
public class User {
    …
    public User( Integer id, String name) {
        this.id =id;
        this.name =name;
    }
    …
}
```

通过自身类型反射匹配参数的配置如下：

```
<bean id="user" class="entity4.User">
  <constructor-arg>
     <value>1</value>
  </constructor-arg >
  <constructor-arg>
     <value>Mike</value>
  </constructor-arg >
</bean>
```

5. 循环依赖问题和工厂方法注入

如果两个 Bean 互相依赖，采用构造函数注入会产出死循环。这时需要将构造函数注入改为 setter 注入。

工厂方法是一个常用的设计模式。在 Spring 中基本上用不到工厂方法注入，有可能在一些遗留系统中会碰到工厂方法。此时用户可以使用 Spring 的工厂方法注入进行配置。篇幅所限，在此不予介绍。

11.4 注入参数详解

Spring 配置文件中可以配置为 Bean 注入字面值，还可以注入其他 Bean、集合元素、null 值。下面以 setter 注入方式为例进行说明。

11.4.1 字面值注入

字面值指用字符串表示的值。字面值使用<value>元素进行注入。默认情况下，基本数据类型及其包装类、String 等类型都可以使用字面值注入方式。Spring 容器会将这些字面值转换为 Bean 类的属性类型。前面的 ch11-1 项目中的 User 类的配置就是使用字面值注入方式，其关键代码如下：

```
<bean id="user" class="entity.User">
   <property name="id" value="1"/>
   <property name="name" value="Mike"/>
   <property name="password" value="123"/>
   <property name="age" value="23"/>
</bean>
```

注入值 value 含有特殊字符，如 &、<、>、"、'时，可以使用<!CD[包含特殊字符的字面值]>，让 XML 解析器将字符串当作普通文本处理；或者使用 XML 转义字符。

例如，如下 Bean user 配置 name 和 password 属性值均含有特殊字符 &，需要分别使用<![CDATA[张 & 三]]>和 &123 对特殊字符 & 进行处理。

```
<bean id="user" class="entity.literal.User">
   <property name="id">
      <value>1</value>
   </property>
   <property name="name">
      <value><![CDATA[张 & 三]]></value>
   </property>
   <property name="password">
      <value>&amp;123</value>
   </property>
</bean>
```

11.4.2 引用其他 Bean

一个 Bean 可以引用其他 Bean。如下 Bean UserService 引用了另外一个 Bean UserDao。

```
package refOtherBean;
public class UserService {
    private UserDao userDao;
    public void setUserDao(UserDao userDao) {
        this.userDao =userDao;
    }
```

```
    // 业务方法:增加用户
    public void add(User user) {
        userDao.add(user);
    }
```

Bean UserService 引用 Bean UserDao 配置如下：

```
<bean id="userService" class="refOtherBean.UserService">
    <property name="userDao">
        <ref bean="userDao"/>
    </property>
</bean>
<bean id="userDao" class=" refOtherBean.UserDao" />
```

<ref>元素可以使用以下 3 个元素引用容器中的其他 Bean。

- bean 属性：引用同一容器或者父容器中的 Bean。这是最常用的方式。
- local 属性：只能引用同一配置文件中定义的 Bean。
- parent 属性：只能引入父类容器中的 Bean。

11.4.3 嵌套 Bean 注入

当 Spring 容器中的某个 Bean A 只会被 Bean B 引用，而不会被容器中任何其他 Bean 引用的时候，则可以将 Bean A 以内部 Bean 的方式注入 Bean B 中。

例如，userDao 仅能被 userService 引用的配置如下：

```
<bean id="userService" class="nestBean.UserService">
    <property name="userDao">
        <bean id="userDao" class=" nestBean.UserDao"/>
    </property>
</bean>
```

11.4.4 null 值注入

如果需要为 Bean 的属性注入一个 null 值，需要使用<null/>。

例如，为 user Bean 的 name 属性注入 null 值的配置如下：

```
<bean id="user" class="nullValue.User">
    <property name="name">
        <null/>
    </property>
</bean>
```

11.4.5 级联属性注入

级联属性注入是指在定义 Bean A 时直接为 Bean B 的属性提供注入值。

例如，Person 类中有属性 phone，phone 所属的 Phone 类有属性 phoneNumber。Person 类的关键代码如下：

```
package cascade;
public class Person {
    private Phone phone;
    public Phone getPhone() {
        return phone;
    }
    public void setPhone(Phone phone) {
        this.phone =phone;
    }
}
```

Phone 类的关键代码如下：

```
package cascade;
public class Phone {
    private String phoneNumber;
    public String getPhoneNumber() {
        return phoneNumber;
    }
    public void setPhoneNumber(String phoneNumber) {
        this.phoneNumber =phoneNumber;
    }
}
```

为 person 的级联属性 phone. phoneNumber 注入值 123456 的配置如下：

```
<bean id="person" class="cascade.Person">
  <property name="phone">
    <ref bean="phone"/>
  </property>
  <property name="phone.phoneNumber" value="123456"/>
</bean>
<bean id="phone" class="cascade.Phone"/>
```

11.4.6 集合注入

Java 中的集合相关的类和接口位于 java. util 包中，主要包括 List、Set、Map、

Properties。Spring 中配置集合属性需要使用专门的配置标签。

1. 注入 List

例如,Person 类中有一个 List 对象 favorites。

```
public class Person {
    private List favorites =new ArrayList();
    public List getFavorites() {
        return favorites;
    }
    public void setFavorites(List favorites) {
        this.favorites =favorites;
    }
}
```

在配置文件中使用 List 标签注入属性值的配置如下:

```
<bean id="person" class="collections.Person">
    <property name="favorites">
        <list>
            <value>唱歌</value>
            <value>运动</value>
            <value>读书</value>
        </list>
    </property>
</bean>
```

数组格式的属性值也可以使用 List 来进行注入;<list>除了使用<value>注入字符串,还可以使用<ref>引用其他 Bean。

2. 注入 Set

例如,Person 类中有一个 Set 对象 favorites。

```
public class Person {
    private Set favorites;
    public Set getFavorites() {
        return favorites;
    }
    public void setFavorites(Set favorites) {
        this.favorites =favorites;
    }
}
```

注入 Set 对象 favorites 配置方式如下:

```
<bean id="person" class="collections.set.Person">
```

```
    <property name="favorites">
        <set>
            <value>运动</value>
            <value>读书</value>
        </set>
    </property>
</bean>
```

3. 注入 Map

例如,Person 类中有一个 Map 对象 favorites。

```
public class Person {
    private Map favorites;
    public Map getFavorites() {
        return favorites;
    }
    public void setFavorites(Map favorites) {
        this.favorites = favorites;
    }
}
```

注入 Map 对象 favorites 配置如下:

```
<bean id="person" class="entity.collections.map.Person">
    <property name="favorites">
        <map>
            <entry>
                <key>
                    <value>key01</value>
                </key>
                <value>唱歌</value>
            </entry>
            <entry>
                <key>
                    <value>key02</value>
                </key>
                <value>跳舞</value>
            </entry>
        </map>
    </property>
</bean>
```

4. 注入 Properties

例如,Person 类中有一个 Properties 对象 favorites。

```
public class Person {
    private Properties favorites;
    public Properties getFavorites() {
        return favorites;
    }
    public void setFavorites(Properties favorites) {
        this.favorites =favorites;
    }
}
```

注入 Properties 对象 favorites 配置如下：

```
<bean id="person" class="collections.properties.Person">
   <property name="favorites">
      <props>
         <prop key="p01">唱歌</prop>
         <prop key="p02">运动</prop>
         <prop key="p01">读书</prop>
      </props>
   </property>
</bean>
```

5. 注入强类型集合

JDK 5.0 提供强类型集合功能，允许为集合元素指定类型。下面 Person 类中的 favorites 是强类型的 Map 类型，其 key 为 Integer，value 为 String。

```
public class Person {
    private Map<Integer, String>favorites;
    public Map<Integer, String>getFavorites() {
        return favorites;
    }
    public void setFavorites(Map<Integer, String>favorites) {
        this.favorites =favorites;
    }
}
```

注入强类型集合 favorites 配置如下：

```
<bean id="person" class="collections. explicittype.Person">
   <property name="favorites">
      <map>
         <entry>
            <key>
               <value>1</value>
            </key>
```

```
                <value>唱歌</value>
            </entry>
            <entry>
                <key>
                  <value>2</value>
                </key>
                <value>跳舞</value>
            </entry>
        </map>
    </property>
</bean>
```

Spring 在注入强类型集合时会判断元素类型并将值转换为相应的数据类型。例如，以上代码中会将<value>1</value>、<value>2</value>转换为 Integer 类型。

6. 集合合并

集合合并允许子<bean>继承父<bean>的同名属性的集合元素，并将子<bean>中配置的集合属性值和父<bean>中配置的同名属性值合并，作为最终<bean>的属性值。

例如，有一个 Person 类代码如下：

```
public class Person {
    private Set favorites;
    public Set getFavorites() {
        return favorites;
    }
    public void setFavorites(Set favorites) {
        this.favorites = favorites;
    }
}
```

父、子 Bean 的配置如下：

```
<bean id="person" abstract="true" class="collections.union.Person">
    <property name="favorites">
        <set>
            <value>运动</value>
        </set>
    </property>
</bean>
<bean id="student" parent="person">
    <property name="favorites">
        <set merge="true">
            <value>旅游</value>
        </set>
```

```
    </property>
</bean>
```

merge="true"表示子 Bean 和父 Bean 中的同名集合合并，merge 取值为 false 则不合并。

merge="true"时，子 Bean student 中的集合 favorites 将与父 Bean person 中的集合 favorites 合并。子 Bean student 中的集合 favorites 拥有两个元素：运动和旅游。

如果是 merge="false"，子 Bean student 中的 favorites 不会与父 Bean 同名集合合并，即子 Bean student 中的集合 favorites 拥有一个元素：旅游。

7. 通过 util 命名空间配置集合类型的 Bean

当需要配置一个集合类型的 Bean，而不是一个集合类型的属性的时候，可以通过 util 命名空间进行配置。

在 beans. xml 配置文件中配置集合类型的 Bean。

```
<?xml version="1.0" encoding="UTF-8"?>
<beans xmlns="http://www.springframework.org/schema/beans"
   xmlns:xsi="http://www.w3.org/2001/XMLSchema-instance"
   xmlns:util="http://www.springframework.org/schema/util"
   xsi:schemaLocation="http://www.springframework.org/schema/beans
   http://www.springframework.org/schema/beans/spring-beans-4.0.xsd
   http://www.springframework.org/schema/util
   http://www.springframework.org/schema/util/spring-util-4.0.xsd">

<util:list id="favoriteList1" list-class="java.util.LinkedList">
   <value>看报</value>
   <value>赛车</value>
</util:list>
<util:set id="favoriteSet1">
   <value>看报</value>
   <value>赛车</value>
</util:set>
<util:map id="favoriteMap1">
   <entry key="01" value="看报"/>
   <entry key="02" value="赛车"/>
</util:map>

<bean id="person" class="collections.util.Person">
  <property name="favorites">
     <ref bean="favoriteSet1"/>
   </property>
 </bean>
</beans>
```

<util:list>标签的 list-class 属性指定了 List 的实现类是 java. util. LinkedList。集

合 Bean Set 和 Map 没有指定实现类，那么其实现类是 Spring 的默认实现类。

定义一个使用集合类型 Bean favoriteSet1 的类 Person，其关键代码如下：

```
public class Person {
    private Set favorites;
    public Set getFavorites() {
        return favorites;
    }
    public void setFavorites(Set favorites) {
        this.favorites = favorites;
    }
}
```

11.5 简化配置

前面使用 XML 配置 Bean 时采用了完整格式的配置方式。Spring 为字面值、引用 Bean 和集合都提供了简化的配置方式。字面值属性简化配置示例见表 11-1。

表 11-1 字面值属性简化配置示例

配置项	简化前	简化后
字面值	<property name="brand"> <value>奔驰</value> </property>	<property name="brand" value="奔驰"/>
构造方法参数	<constructor-arg type="String"> <value>奔驰</value> </constructor>	<constructor-arg type="String" value="奔驰"/>
集合元素	<map> <entry> <key> <value>01</value> </key> <value>看报</value> </entry> </map>	<map> <entry key="01" value="看报"/> </map>

引用其他 Bean 简化配置示例见表 11-2。

表 11-2 引用其他 Bean 简化配置

配置项	简化前	简化后
字面值	<property name="userDao"> <ref bean="userDao"></ref> </property>	<property name="userDao" ref="userDao"/>

续表

配置项	简化前	简化后
构造方法参数	<constructor-arg type="dao. UserDao"> <ref bean="userDao"/> </constructor>	<constructor-arg ref="userDao"/>
集合元素	<map> <entry> <key> <ref bean="keyBean"/> </key> <ref bean="valueBean"/> </entry> </map>	<map> <entry key-ref="keyBean"/> <value-ref="valueBean"/> </map>

11.6 Bean的作用域和生命周期

11.6.1 Bean的作用域

配置 Bean 时需要指定 Bean 的作用域。Spring 支持的 Bean 作用域类型说明见表 11-3。

表 11-3 Bean 的作用域类型说明

作用域类型	作用域说明
singleton	每个 Spring IOC 容器只有一个 Bean 实例。 这是 Spring Bean 的默认作用域
prototype	每次使用一个 Bean 时创建一个 Bean 实例
request	每个 HTTP 请求会创建一个 Bean 实例。 该作用域仅适用于 Web 应用环境
session	每个 HTTP Session 会创建一个 Bean 实例。 该作用域仅适用于 Web 应用环境
global-session	该作用域将 bean 的定义为全局 HTTP Session， 一般用于 Portlet 应用环境

Bean 的作用域还可以是用户自定义类型。自定义作用域在实际应用中较少使用，在此不进行介绍。

指定 Bean 作用域使用 scope = " singleton | prototype | request | session | global-session"。以下代码指定 Bean 的作用域为 singleton：

```
<bean id="user" class="entity.User" scope="singleton">
    …
```

```
</bean>
```

11.6.2 Bean的生命周期

Spring 容器中 Bean 的生命周期如下。

(1) Spring 容器对 Bean 实例化;

(2) Spring 将值或引用 Bean 注入 Bean 对应的属性中;

(3) 如果 Bean 实现了 BeanNameAware 接口,那么 Spring 将 Bean 的 ID 传递给 setBeanName()方法;

(4) 如果 Bean 实现了 BeanFactoryAware 接口,那么 Spring 将调用 setBeanFactory()方法传入 BeanFactory 容器实例;

(5) 如果 Bean 实现了 ApplicationContextAware 接口,那么 Spring 将调用 setApplicationContext()方法传入 Bean 所在的应用上下文的引用;

(6) 如果 Bean 实现了 BeanPostProcessor 接口,那么 Spring 将调用 postProcessBeforeInitialization()方法;

(7) 如果 Bean 实现了 InitializingBean 接口,那么 Spring 将调用 afterPropertiesSet()方法。如果 Bean 使用了 init-method 声明了初始化方法,那么该方法会被调用;

(8) 如果 Bean 实现了 BeanPostProcessor 接口,那么 Spring 将调用 postProcessAfterInitialization()方法;

(9) 此时 Bean 已经准备就绪,可以被应用程序使用了。Bean 将一直存在于应用上下文中,直到该应用上下文销毁;

(10) 如果 Bean 实现了 DisposableBean 接口,那么 Spring 将调用 destroy()方法。如果 Bean 使用 destroy-method 声明了销毁方法,那么该方法会被调用。

Bean 生命周期回调方法是指容器创建 Bean 后调用 Bean 的某个方法进行初始化和销毁工作。在 Bean 上定义初始化和销毁的方法,例如 init()和 destroy();在 beans.xml 配置＜bean init-method="init" destroy-method="destroy"＞。

11.7 使用 XML 的自动装配

首先回顾手动装配一个 Bean 的过程。如下代码是手动装配一个 Bean userDao。

```
<bean id="userService" class="spring.service.UserService"autowire="no">
  <property name="userDao">
     <ref bean="userDao"/>
  </property>
</bean>
<bean id="userDao" class="spring.dao.UserDao"/>
```

默认 autowire 属性值为 no,即不使用自动装配,需要通过＜ref＞标签手动装配 Bean

userDao。

自动装配使用方式：

<bean id="beanId" class="package. SomeBean" autowire="自动装配类型">。

Spring 提供了 4 种自动装配类型。

(1) byName：把与 Bean 的属性具有相同名字的其他的 Bean 自动装配到 Bean 对应的属性当中。如果没有与属性名字相匹配的 Bean，则该属性不进行装配。例如，UserService 有个名为 userDao 的属性，如果容器中存在一个 userDao Bean，那么 userDao Bean 会被自动装配给 UserService 的 userDao。

(2) byType：把与 Bean 的属性具有相同类型的其他 Bean 自动装配到 Bean 的对应属性当中。如果没有根属性类型的 Bean，则属性不被装配。例如，UserService 有个类型为 UserDao 的属性，如果容器中存在一个类型为 UserDao Bean，那么 UserDao Bean 会被自动装配给 UserService 的 UserDao 属性。

(3) constructor：把与 Bean 的构造函数参数具有相同类型的其他 Bean 自动装配到 Bean 的构造器对应的参数中。

(4) autodetect：首先尝试使用 constructor 进行自动装配，如果失败，则尝试使用 byType 进行自动装配。

另外，通过设置配置文件根元素<beans>中的 default-autowire 属性可以配置全局自动装配。default-autowire 的默认值为 no，表示不启用自动装配。<beans>中定义的自动装配策略可以被<bean>的自动装配策略覆盖。

使用 XML 配置时启用自动装配的功能可以减少配置文件的代码量，但是在大型项目中不推荐使用，因为容易引起混乱。基于注解的配置方式默认采用的是 byType 自动装配类型。

11.8 使用 Java 配置

Spring 配置 Bean 包括使用 XML 配置、使用 Java 配置、混合配置(同时使用 XML 和 Java 配置)。前面配置 Bean 都是使用 XML 配置 Bean。

使用 Java 配置包括使用 Java 手动配置、使用 Java 自动装配。

11.8.1 使用 Java 手动配置

1. Bean 属性值注入

使用 Java 配置为 User Bean 注入属性值。User 类的关键代码如下：

```
package config;
public class User {
    private int id;
```

```
    private String name;
    …
}
```

使用 Java 配置的配置类 Config 代码如下：

```
package config;
…
@Configuration
public class Config {
    @Bean
    public User user() {
        User u =new User();
        u.setId(1);
        u.setName("Mike");
        return u;
    }
}
```

类前的@Configuration 注解说明 Config 类是配置类。

user()方法返回一个 User 类型的对象。@Bean 注解说明返回的 User 对象将被注册为 Spring 应用上下文中的一个 Bean。方法名 user 将作为该 Bean 的 ID。创建的 User 对象 u 调用 setter 方法设置 id 和 name 属性表明是采用了 setter 注入。

以上代码与以下的 XML 配置是等效的。

```
<bean id="user" class="config.User" />
```

也可以采用构造方法注入。使用 Java 配置的构造方法注入要求 Bean 必须有对应的构造方法。例如，User 类如果采用构造方法注入，其关键代码如下：

```
package config;
public class User {
    …
    public User(int id, String name) {
        super();
        this.id =id;
        this.name =name;
    }
}
```

采用构造方法注入时 Java 配置类 Config 关键代码如下：

```
package config0;
…
@Configuration
public class Config {
```

```
    @Bean
    public User user() {
        User u =new User(1, "构造方法注入");
        return u;
    }
}
```

采用 setter 注入的测试类 DITest 代码如下：

```
package config;
…
public class DITest {
    public static void main(String[] args) {
        ApplicationContext annotationContext
            =new AnnotationConfigApplicationContext("Config");
        User user =annotationContext.getBean("user", User.class);
        System.out.println(user.getId() +"-" +user.getName());
    }
}
```

AnnotationConfigApplicationContext 可以实现基于 Java 的配置类加载 Spring 的应用上下文。

2. 注入依赖对象

通常应用中的对象有依赖关系。常见 Java Web 程序开发中,DAO 层对象负责数据访问,Service 层负责业务逻辑处理。Service 层会调用 DAO 层对象即 Service 层对象依赖 DAO 层对象。下面使用 UserService 和 UserDao 展示将 UserDao 对象注入 UserService 中。

User 类的关键代码如下。

```
package config1;
public class User {
    private int id;
    private String name;
    …
}
```

UserDao 接口代码如下：

```
package config1;
public interface UserDao {
    public void add(User user);
}
```

UserDaoImpl 类代码如下：

```
package config1;
public class UserDaoImpl implements UserDao {
    @Override
    public void add(User user) {
        System.out.println("增加了一个用户：" +user.getName());
    }
}
```

UserService 接口代码如下：

```
package config1;
public interface UserService {
    public void add(User user);
}
```

UserServiceImpl 类代码如下：

```
package config1;
public class UserServiceImpl implements UserService {
    private UserDao userDao;
    @Override
    public void add(User user) {
        userDao.add(user);
    }
    public void setUserDao(UserDao userDao) {
        this.userDao =userDao;
    }
}
```

使用 Java 配置的配置类 Config 的关键代码如下：

```
package config1;
…
@Configuration
public class Config {
    @Bean
    public UserDao userDao() {
        return new UserDaoImpl();
    }
    @Bean//(name="myservice")
    public UserService userService(UserDao userDao) {
        UserServiceImpl userService =new UserServiceImpl();
        userService.setUserDao(userDao);
        return userService;
    }
}
```

在 Java 配置类 Config 中，通过@Bean 定义了两个方法：userDao 和 userService，默认情况下方法名作为 Bean 的 id 即 Spring 会创建两个 Bean，其 id 分别为 userDao 和 userService。如果想改变 Bean 的名称，可以使用@Bean 的属性 name，如上述代码被注释掉的(name="myservice")。userService 方法中使用 setter 注入方式注入了依赖对象 userDao，因此 UserServiceImpl 实现类中必须有 setUserDao(UserDao userDao)方法。

测试类 DITest 代码如下：

```
package config1;
import org.springframework.context.annotation.AnnotationConfigApplicationContext;
public class DITest {
    public static void main(String[] args) {
        AnnotationConfigApplicationContext context =
            new AnnotationConfigApplicationContext(Config.class);
        UserService userService = (UserService) context.getBean("userService");
        User user =new User();
        user.setId(1);
        user.setName("Config2");
        userService.add(user);
        context.close();
    }
}
```

在测试类 DITest 中，通过 getBean 方法获取 userService Bean，实现了增加一个用户的模拟操作。

11.8.2 使用 Java 自动装配

此处还是以 11.8.1 节的案例程序说明基于 Java 自动装配。UserDao 接口、UserService 接口、测试类 DITest 代码不变。

实现类 UserDaoImpl 和 UserServiceImpl 均在类前使用@Component 注解。@Component 注解表示这是一个组件类并由 Spring 把该类实例化到容器中。

Spring 还提供了 3 个功能与@Component 等效的注解：@Repository、@Service、@Controller，分别用在 DAO 层实现类、Service 层实现类、Web 层的控制器实现类前。这 3 个注解与@Component 效果一样。提供以上这 3 个特殊注解主要是为了让类本身的用途更加清晰，因此在 Web 层、Service 层、DAO 层推荐使用这 3 个注解。

UserDaoImpl 的关键代码如下：

```
@Component
public class UserDaoImpl implements UserDao {
    ...
}
```

UserServiceImpl 的关键代码如下：

```java
package config2;
@Component("userService")
public class UserServiceImpl implements UserService {
    private UserDao userDao;
    @Autowired // (required =true)
    public UserServiceImpl(UserDao userDao) {
        this.userDao =userDao;
    }
    @Override
    public void add(User user) {
        userDao.add(user);
    }
}
```

在 UserServiceImpl 类的构造方法上使用@Autowired 注解为该类注入 UserDao userDao。@Autowired 注解还可以用在类属性上或是 setter 方法上或是其他任意方法上。

Java 配置类 Config 在类前需要使用@ComponentScan 注解，用于开启组件自动扫描。其关键代码如下：

```java
@Configuration
@ComponentScan
public class Config {
    @Bean
    public UserDao userDao() {
        return new UserDaoImpl();
    }
    @Bean // (name="myservice")
    public UserService userService(UserDao userDao) {
        UserServiceImpl userService =new UserServiceImpl(userDao);
        return userService;
    }
}
```

@ComponentScan 注解如果没有任何属性设置，则默认会扫描 Java 配置类所在的包及其子包，并为带@Component 注解的类在 Spring 应用上下文中创建 Bean 对象。本例中 Spring 会扫描 Config 类所在的包即 config 包及其子包下所有带@Component 的类，为这些类在 Spring 应用上下文中创建 Bean 对象。

如果需要显式地指定扫描包，可以采用如下形式：

```java
@ComponentScan("config")
```

或其等价形式：

```java
@ComponentScan(basePackages="config")
```

表示自动扫描 config 包及其子包，并为带@Component 注解的类在 Spring 应用上下文中创建 Bean 对象。

可以显式地指定多个包：

```
@ComponentScan(basePackages={"dao","service"})
```

表示自动扫描 dao 和 service 包及其子包，并为带@Component 注解的类在 Spring 应用上下文中创建 Bean 对象。

还可以配置@ComponentScan 注解的 basePackageClasses 属性自动扫描指定的类所在包及其子包：

```
@ComponentScan(basePackagClasses={UserDao.class, UserService.class})
```

表示自动扫描 UserDao、UserService 所在的包及其子包，并为带@Component 注解的类在 Spring 应用上下文中创建 Bean 对象。

如果在 Spring 的 XML 配置文件中启用自动组件扫描，则需要使用 Spring 的 context 命名空间以及<context：component-scan>元素。如下代码使用<context：component-scan>元素的 base-package＝"config"，指定自动扫描 config 包及其子包，并为带@Component 注解的类在 Spring 应用上下文中创建 Bean 对象。

```
<beans xmlns="http://www.springframework.org/schema/beans"
    xmlns:xsi="http://www.w3.org/2001/XMLSchema-instance"
    xmlns:context="http://www.springframework.org/schema/context"
    xsi:schemaLocation="http://www.springframework.org/schema/beans
    http://www.springframework.org/schema/beans/spring-beans-2.5.xsd
    http://www.springframework.org/schema/context
    http://www.springframework.org/schema/context/spring-context-2.5.xsd">
    <context:component-scan base-package="config" />
</beans>
```

11.9 AOP

11.9.1 AOP 简介

AOP(Aspect Oriented Programing)通常翻译为面向切面的编程，是 OOP 的补充。借助 AOP，可以实现分离应用的业务逻辑与横切关注点。

通常在某个应用的多个业务类中存在一些非业务逻辑的代码，如日志记录、性能监控、事务管理、访问控制等。这些散落在多个类中的非业务代码被称为横切关注点。

如下是一个用户服务模块的部分代码段：

```
public class UserServiceImpl{
    public void add(User user){
```

```
        pmomitor.start();
        transManager.beginTransaction();
        userDao.add(User user);
        transManager.commit();
        pmomitor.end();
    }
    public void delete(int id){
        pmomitor.start();
        transManager.beginTransaction();
        userDao.delete(int id);
        transManager.commit();
        pmomitor.end();
    }
    ...
}
```

以上代码中，核心业务逻辑是调用 DAO 层对象 userDao 的 add(User user)和 delete(int id)方法增加和删除一个用户。然而在增加和删除用户前后，存在性能监控、事务访问等非业务逻辑代码(横切关注点)。AOP 希望将这些散落在各个业务逻辑中的相同代码通过横切方式抽取到一个模块中。

横切关注点可以被模块化为特殊的类，这些类被称为切面。这样做有两个好处：一是每个关注点集中在一个地方而不是散落在多处代码中；二是服务模块更加简洁。因为服务模块只包含核心功能的代码，而次要关注点的代码被转移到切面中了。

一些常用的基于 Java 的 AOP 实现如下。

(1) AspectJ。

AspectJ 是目前最完善的 AOP 语言，对 Java 编程语言进行了扩展，定义了 AOP 语法，能够在编译期提供横切代码的织入。AspectJ 提供了两种横切实现机制：一种称为动态横切(Dynamic Crosscuting)，另一种称为静态横切(Static Crosscuting)。

(2) AspectWerkz。

AspectWerkz 是基于 Java 的简单、动态和轻量级的 AOP 框架，支持运行期或者类装载期织入横切代码，它拥有一个特殊的类装载器。它已经与 AspectJ 项目合并，第一个发行版是 AspectJ5。扩展 AspectJ 语言，以基于注解的方式支持类似 AspectJ 的代码风格。

(3) JBoss AOP。

JBoss 是一个开源的符合 Java EE 规范的应用服务器，作为 Java EE 规范的补充，JBoss 中引入了 AOP 框架，为普通 Java 应用提供了 Java EE 服务，而无须遵循 EJB 规范。JBoss 通过类加载时使用 Javassis 对字节码操作实现动态 AOP 框架。

(4) Spring AOP。

Spring AOP 使用纯 Java 实现，不需要专门的编译过程和特殊的类加载器，在运行期通过代理方式向目标类织入增强代码。Spring 并不提供最完整的 AOP 实现，主要侧重一种与 Spring IoC 容器整合的 AOP 实现，以解决企业级开发中的常见问题。

11.9.2 AOP的术语

1. 连接点

程序执行的某个特定位置，比如类初始化前、初始化后，方法调用前、方法调用后等。

2. 切点

每个类都拥有多个连接点，但是可能只关心其中的几个连接点。AOP通过切点来定位特定的连接点。例如，某个类有两个方法，这两个方法都是连接点，可能只需要其中的一个方法，可以使用切点定位需要的方法。

3. 增强

织入到目标类连接点上的一段程序代码。

4. 目标对象

增强逻辑的织入目标类。

5. 引介

引介是一种特殊的增强，它为类添加一些属性和方法。

6. 织入

将增强添加到目标类的具体连接点上的过程。AOP有3种织入方式：

- 编译期织入。要求使用特殊的Java编译器。
- 类装载织入。要求使用特殊的类装载器。
- 动态代理织入。在运行期为目标类添加增强生成子类的方式。

7. 代理

一个类被AOP织入增强后会产生一个结果类，这个融合了原类和增强逻辑的类称为代理类。代理类的对象称为代理对象。

8. 切面

由切点和增强组成，既包括了横切逻辑的定义，也包括了连接点的定义。

11.9.3 Spring AOP基础

Spring AOP仅支持方法连接点，即仅能在方法调用前、方法调用后、方法抛出异常时执行织入增强。

1. Spring 增强的类型

按照在目标类方法中的连接点位置，Spring 增强分为 5 类。

- 前置增强：表示在目标方法调用之前执行增强。
- 后置增强：表示在目标方法调用之后执行增强。
- 环绕增强：表示在目标方法调用之前和之后执行增强。
- 异常增强：表示在目标方法调用发生异常之后执行增强。但是如果目标方法中捕获了异常，那么异常增强不会执行。
- 引介增强：表示在目标类中添加一些新的方法和属性。

2. Spring 对 AOP 的支持

Spring 对 AOP 的支持分 4 类：基于代理的经典 Spring AOP、纯 POJO 切面、@AspectJ注解驱动的切面、注入式 AspectJ 切面。其中前 3 种都是 Spring AOP 实现的不同形式。Spring 经典的 AOP 直接使用 FactoryBean，显得笨重和复杂，现在很少使用；纯 POJO 切面是通过在 XML 中使用 Spring 的 AOP 命名空间将 POJO 转化为切面；@AspectJ注解驱动的切面是通过注解配置实现 AOP；注入式 AspectJ 切面通常用于 Spring 的方法连接点不能满足要求的情况，例如，需要在构造方法或属性上织入增强。

3. Spring AOP 原理

Spring 底层使用 JDK 或 CGLib 动态代理技术为目标织入横切逻辑。Spring 在运行期通过在代理类中包裹切面、将切面织入到 Spring 管理的 Bean 中。代理类封装了目标类并拦截被增强的方法调用，再将调用转发给真正的目标 Bean。当拦截到方法调用时，在调用目标 Bean 方法之前，代理会执行切面逻辑。Spring 运行时才创建代理对象，所以不需要特殊的编译器织入 Spring AOP 的切面。

4. AspectJ 切点表达式

Spring AOP 使用 AspectJ 的切点表达式语言来定义切点。Spring 仅支持 AspectJ 切点指示器的一个子集。Spring AOP 所支持的 AspectJ 切点指示器见表 11-4。

表 11-4 Spring AOP 所支持的 AspectJ 切点指示器

AspectJ 指示器	说　明
arg()	限制连接点匹配参数为指定类型的执行方法
@arg()	限制连接点匹配参数由指定注解标注的执行方法
execution()	用于匹配是连接点的执行方法
this()	限制连接点匹配 AOP 代理的 Bean 引用为指定类型的类
target()	限制连接点匹配目标对象为指定类型的类
@target()	限制连接点匹配待定的执行对象，这些对象对应的类要具备指定类型的注解

续表

AspectJ 指示器	说　明
within()	限制连接点匹配指定的类型
@within()	限制连接点匹配指定注解所标注的类型(当使用 Spring AOP 时,方法定义在由指定的注解所标注的类里)
@annotaion	限制匹配带有指定注解连接点

在上述 Spring 支持的 AspectJ 切点指示器中,只有 execution 指示器时执行匹配,其他指示器都是用于限制匹配。

例如,如下切点表达式表示在执行 AOP 包中的 UserService 的 add()方法时触发增强:

```
execution(* aop.UserService.add(..))
```

- 第一个 * 表示方法返回值为任意。
- add(..)表示选择任意的 add()方法,与 add()方法的参数无关。

指示器之间可以混合使用,用 and、or、not(&&、||、!)连接。

例如,如下切点表达式使用 within 限制仅仅匹配 AOP 包:

```
execution(* aop.UserService.add(..)) && within(aop.*)
```

5. AspectJ 定义增强的注解

AspectJ 提供了 5 个注解来定义增强,见表 11-5。

表 11-5　AspectJ 定义增强的 5 个注解

注　解	说　明
@Before	使用在方法之前,表示该方法是前置增强,即该方法在目标方法调用之前执行
@After	使用在方法之后,表示该方法是后置增强,即该方法在目标方法调用之后执行
@Around	使用在方法之前,表示该方法是环绕增强,即该方法在目标方法调用之前和之后执行
@AfterReturning	使用在方法之前,表示该方法在目标方法返回之后执行。
@AfterThrowing	使用在方法之前,表示该方法是异常增强,即该方法在目标方法抛出异常后执行

11.9.4　使用注解实现 Spring AOP 前置和后置增强

在一般的 Web 应用中,服务层调用数据访问层;数据访问层通过借助于持久化 API(如 JPA、Hibernate、JDBC)实现数据访问。

下面介绍使用注解配置实现 Spring AOP,在 UserService 的 add(User user)方法之前和之后增加日志记录功能。涉及实体类 User,数据访问层的 UserDao 接口及其实现类

UserDaoImpl、服务层 UserService 接口及其实现类 UserServiceImpl。

User 类的关键代码如下：

```
package aop;
public class User {
    private int id;
    private String name;
    public User(int id, String name) {
        this.id = id;
        this.name = name;
    }
    …
}
```

UserDao 接口的关键代码如下：

```
package aop;
public interface UserDao {
    public void add(User user);
}
```

UserDaoImpl 类实现了 UserDao 接口中存在模拟增加用户的 add(User user)方法，其关键代码如下：

```
package aop;
import org.springframework.stereotype.Component;
@Component
public class UserDaoImpl implements UserDao {
    @Override
    public void add(User user) {
        System.out.println("增加了一个用户：" + user.getName());
    }
}
```

UserService 接口代码如下：

```
package aop;
public interface UserService {
    public void add(User user);
}
```

UserServiceImpl 类实现了 UserService 接口，其关键的代码如下：

```
package aop;
…
@Component("userService")
public class UserServiceImpl implements UserService {
    private UserDao userDao;
```

```
    @Autowired
    public UserServiceImpl(UserDao userDao) {
        this.userDao =userDao;
    }
    @Override
    public void add(User user) {
        userDao.add(user);
    }
}
```

1. 定义切面

使用注解来创建切面是AspectJ 5所引入的关键特性。AspectJ 5面向注解的模型通过少量注解把任意类转变为切面。切面类Log代码如下：

```
package aop;
...
@Aspect
public class Log {
    @Before("execution(* * aop.UserService.add(..))")
    public void logBefore() {
        System.out.println("业务方法执行之前--记录日志");
    }
    @After("execution(* * aop.UserService.add(..))")
    public void logAfter() {
        System.out.println("业务方法执行之后--记录日志");
    }
}
```

Log类前的注解@Aspect表示该类是一个切面。该类中的两个方法logBefore()和logAfter()前面分别使用了

```
@Before("execution(* * aop.UserService.add(..))")
```

和

```
@After("execution(* * aop.UserService.add(..))")
```

表示logBefore()和logAfter()方法将用在aop.UserService.add()方法之前和之后被调用。

2. 使用Java配置Bean和启用AspectJ注解自动代理

基于Java配置的类Config中需要配置一个Bean Log和启用AspectJ注解自动代理，其代码如下：

```
package aop;
```

```
...
@Configuration
@EnableAspectJAutoProxy
@ComponentScan
public class Config {
    @Bean
    public Log log() {
        return new Log();
    }
}
```

@EnableAspectJAutoProxy 在 Config 类前使用，表示启用 AspectJ 自动代理。

Config 类中的 log()方法前使用@Bean，表示该类配置为一个 Spring Bean。

测试类 DITest 代码如下：

```
package aop;
import org.springframework.context.annotation.AnnotationConfigApplicationContext;
/*
 *
 * 完全使用注解、自动扫描和自动注入来配置 Bean 和 AOP
 */
public class DITest {
    public static void main(String[] args) {
        AnnotationConfigApplicationContext context
            =new AnnotationConfigApplicationContext(Config.class);
        UserService userService = (UserService) context.getBean("userService");
        userService.add(new User(1, "aop"));
        context.close();
    }
}
```

DITest 执行结果如下：

```
业务方法执行之前--记录日志
增加了一个用户:aop
业务方法执行之后--记录日志
```

11.9.5 使用注解实现 Spring AOP 环绕增强

下面介绍使用注解实现 Spring AOP 方式来实现环绕增强，在 UserService 的 add(User user)方法之前和之后增加日志记录。

涉及实体类 User、数据访问层的 UserDao 接口及其实现类 UserDaoImpl、服务层 UserService 接口及其实现类 UserServiceImpl、配置类 Config、测试类 DITest、切面类 Log。除了切面类 Log，其他类代码与 11.9.4 节相同。Log 类代码如下：

```
package aop1;
…
@Aspect
public class Log {
    @Around("execution(* * aop1.UserService.add(..))")
    public void logBeforeAndAfter(ProceedingJoinPoint jp) {
        try {
            System.out.println("业务方法执行之前--记录日志");
            jp.proceed();
            System.out.println("业务方法执行之后--记录日志");
        } catch (Throwable e) {
            e.printStackTrace();
        }
    }
}
```

在方法 logBeforeAndAfter(ProceedingJoinPoint jp)前使用@Around("execution(** aop1. UserService. add(..))"),表示该方法会对 UserService 的 add 方法进行环绕增强。

jp. proceed()表示将调用被增强的方法。本例中调用了 UserService 的 add 方法。jp 参数的类型是 ProceedingJoinPoint。

11.9.6 使用 XML 配置 Spring AOP 实现前置和后置增强

Spring AOP 可以基于 Java 配置类方式或基于 XML 文件配置方式。前面的案例都是基于 Java 配置类方式。下面以 Spring AOP 的前置和后置增强的例子为基础,将其改为 XML 配置方式。

涉及实体类 User、数据访问层的 UserDao 接口及其实现类 UserDaoImpl、服务层 UserService 接口及其实现类 UserServiceImpl、配置类 Config、测试类 DITest、切面类 Log。User 类、UserDao 接口、UserService 接口代码不变。实现类 UserDaoImpl 和 UserServiceImpl 中移除了注解@Component,需要在 XML 中配置这两个 Bean。UserDaoImpl 的关键代码如下:

```
package aop2;
public class UserDaoImpl implements UserDao {
    @Override
    public void add(User user) {
        System.out.println("增加了一个用户:" +user.getName());
    }
}
```

UserServiceImpl 的关键代码如下:

```
package aop2;
```

```java
public class UserServiceImpl implements UserService {
    private UserDao userDao;
    public void setUserDao(UserDao userDao) {
        this.userDao = userDao;
    }
    @Override
    public void add(User user) {
        userDao.add(user);
    }
}
```

切面类 Log 移除了相关 AOP 注解，其关键代码如下：

```java
package aop2;
public class Log {
    public void logBefore() {
        System.out.println("业务方法执行之前--记录日志");
    }
    public void logAfter() {
        System.out.println("业务方法执行之后--记录日志");
    }
}
```

XML 配置文件 beans.xml 代码如下：

```xml
<?xml version="1.0" encoding="UTF-8"?>
<beans xmlns="http://www.springframework.org/schema/beans"
  xmlns:xsi="http://www.w3.org/2001/XMLSchema-instance"
  xmlns:aop="http://www.springframework.org/schema/aop"
  xsi:schemaLocation="http://www.springframework.org/schema/beans
  http://www.springframework.org/schema/beans/spring-beans-4.0.xsd
  http://www.springframework.org/schema/aop
  http://www.springframework.org/schema/aop/spring-aop-4.0.xsd ">
  <aop:config>
      <aop:aspect id="log" ref="log">
         <aop:pointcut id="add"
            expression="execution(* aop2.UserService.add(..))"/>
         <aop:before pointcut-ref="add" method="logBefore"/>
         <aop:after pointcut-ref="add" method="logAfter"/>
      </aop:aspect>
  </aop:config>
  <!--Definition for Bean userDao and UserService -->
  <bean id="userDao" class="aop2.UserDaoImpl"/>
  <bean id="userService" class="aop2.UserServiceImpl">
      <property name="userDao">
          <ref bean="userDao"/>
```

```
        </property>
    </bean>
    <!--Definition for Beanlog -->
    <bean id="log" class="aop2.Log"/>
</beans>
```

<bean id="log" class="aop2.Log"/>配置一个 Spring Bean，其 id 为 log，其等价使用注解的代码如下：

```
@Bean
public Log log() {
    return new Log();
}
```

<bean id="userDao" class="aop2.UserDaoImpl"/>使用 XML 配置了 Spring Bean，其 id 为 userDao。

<bean id="userService" class="aop2.UserServiceImpl">使用 XML 配置了 Spring Bean，其 id 为 userService。

XML 配置文件头部的如下代码声明了 Spring AOP 命名空间：

```
xmlns:aop="http://www.springframework.org/schema/aop"
```

<aop:aspect id="log" ref="log">定义切面 log Bean。

<aop:pointcut id="add" expression="execution(* aop2.UserService.add(..))"/>定义切点，其 id 为 add，expression 属性定义切点为 aop2.UserService.add(..)。

<aop:before pointcut-ref="add" method="logBefore"/>定义了前置增强，表示在执行切点 aop2.UserService.add(..)方法之前执行增强。

<aop:before pointcut-ref="add" method="logAfter"/>定义了后置增强，表示在执行切点 aop2.UserService.add(..)方法之后执行增强。

测试类 DITest 的关键代码如下：

```
package aop2;
/*
 *使用 xml 配置 bean 和 aop。
 */
public class DITest {
    public static void main(String[] args) {
        ApplicationContext context =
            new ClassPathXmlApplicationContext("aop2/beans.xml");
        UserService userService = (UserService) context.getBean("userService");
        userService.add(new User(1, "aop2"));
    }
}
```

小　结

本章主要介绍了 Spring 框架的核心基础知识。首先以 Java SE 环境下使用 Spring 为例说明了 Spring 的一般使用步骤。接着讲述了 Spring 核心的 IOC 容器、依赖注入。Spring 配置 Bean 注入属性值分为使用 XML 配置、使用 Java 配置、混合配置，对于前两种进行了详细讨论。最后讲解了 Spring 的另一个核心基础 AOP，并根据 XML 配置和 Java 配置分别进行了讨论。

思考与习题

1. Spring 框架的主要特点是什么？写出 Java SE 环境下使用 Spring 的一般步骤。
2. Spring 提供的两种 IOC 容器分别是什么？二者有何区别？
3. 经常使用的 ApplicationContext 接口的实现主要有哪些？
4. 什么是依赖注入？Spring 依赖注入方式包括哪些？常用的是什么？
5. Employee 类有属性 int id、String name、int age、String sex。请使用 XML 配置的 setter 注入和构造方法注入，为 Employee 类的对象 employee 注入属性值：id 为 1，name 为 Mike，age 为 22，sex 为男。
6. 注入值 value 含有特殊字符如 & 等时如何解决？
7. 假如 UserController 引用了 UserService，请写出 XML 配置代码。
8. 什么是嵌套 Bean 注入？
9. 举例说明如何实现级联属性注入。
10. 举例说明各种集合注入的配置。
11. 对第 6 题和第 8 题使用 XML 简化配置方式进行配置。
12. 写出 Spring 支持的 Bean 作用域类型说明。
13. 举例说明 Bean 生命周期回调方法。
14. 对于使用 XML 自动装配，Spring 提供了几种自动装配类型？分别是什么？
15. 对第 6 题和第 8 题使用 Java 配置方式进行配置。
16. @Component 注解的作用是什么？@Repository、@Service、@Controller 的作用是什么？
17. 什么是 AOP？常用的基于 Java 的 AOP 实现有哪些？
18. 写出 11.9.2 节种的 8 个 AOP 术语。
19. Spring 增强分为几类？分别是什么？
20. Spring 对 AOP 的支持分几类？分别是什么？
21. 写出表示在执行 AOP 包中的 UserAction 的 add()方法时触发增强的切点表达式。
22. AspectJ 提供了 5 个定义增强的注解分别是什么？
23. 对于 11.9.4 节、11.9.5 节、11.9.6 节的案例，请编写完整的程序并进行测试。

第12章

使用 Spring 持久化

Spring 支持对 JDBC、JPA、Hibernate、MyBatis、JDO 等多种持久化技术的集成。Spring 定义了针对持久化的通用异常体系和简化持久化操作的模板技术，对持久化操作提倡面向接口编程。

12.1 使用 Spring JDBC

使用原始的 JDBC 访问数据库需要写大量的代码来进行打开和关闭数据库连接、处理异常、处理事务等底层操作。使用 Spring JDBC 进行数据操作时，Spring JDBC 负责这些底层操作。

12.1.1 使用 JdbcTemplate

Spring JDBC 提供了 JdbcTemplate 模板类用于完成访问数据库操作：执行 SQL 查询、更新语句和存储过程调用；执行迭代结果集和提取返回参数值；捕获 JDBC 异常并转换为 Spring 异常体系结构的相关类。

下面介绍使用 JdbcTemplate 完成对数据库 testDB 中 user 表的增、删、改、查。

首先创建 Eclipse Java 项目 ch12-01，并配置该项目的构建路径添加 Spring4.1.6 的相关 jar 包、commons-logging-1.2.jar 以及 MySQL5 的 JDBC 驱动程序 mysql-connector-java-5.0.6-bin.jar。

1. 配置数据源

此处采用基于 XML 配置数据源。在 Spring 配置文件 beans.xml 中，配置连接数据库相关信息如驱动程序、数据库连接 URL、登录数据库用户名和密码。数据源配置如下：

```
<bean id="dataSource"
    class="org.springframework.jdbc.datasource.DriverManagerDataSource">
    <property name="driverClassName" value="com.mysql.jdbc.Driver"/>
    <property name="url" value="jdbc:mysql://localhost:3306/testDB"/>
```

```
    <property name="username" value="root"/>
    <property name="password" value="root"/>
</bean>
```

DriverManagerDataSource 是 Spring 提供的一个简单的数据源实现类。该类并没有提供池化连接机制，即每次获取一个新的数据库连接而不是从连接池获取连接。该数据源类适合简单的应用或单元测试场景。如果是在生产环境下，建议使用 Spring 支持的第三方开源数据源，如 Apache BasicDataSource 或 C3P0 数据源。

2. 配置 JdbcTemplate

在 Spring 配置文件 beans.xml 中，配置 JdbcTemplate 如下：

```
<bean id="jdbcTemplate" class="org.springframework.jdbc.core.JdbcTemplate">
    <property name ="dataSource" ref="dataSource" />
</bean>
```

在配置 JdbcTemplate 时需要注入前面配置的 dataSource。

3. 启用自动扫描

在 Spring 配置文件 beans.xml 中，自动扫描配置如下：

```
<?xml version="1.0" encoding="UTF-8"?>
<beans xmlns="http://www.springframework.org/schema/beans"
    xmlns:xsi="http://www.w3.org/2001/XMLSchema-instance"
    xmlns:context="http://www.springframework.org/schema/context"
    xmlns:aop="http://www.springframework.org/schema/aop"
    xsi:schemaLocation="http://www.springframework.org/schema/beans
    http://www.springframework.org/schema/beans/spring-beans-4.0.xsd
    http://www.springframework.org/schema/aop
    http://www.springframework.org/schema/aop/spring-aop-4.0.xsd
    http://www.springframework.org/schema/context
    http://www.springframework.org/schema/context/spring-context-4.0.xsd" >
    <context:component-scan base-package="dao,service" />
```

<context:component-scan base-package="dao,service" />表示 Spring 自动扫描 dao 和 service 包中带注解@Component 的类并为其创建 Bean。需要引入 context 命名空间和 context 的 schema 路径，见上述加粗部分代码。

4. 使用 JdbcTemplate 实现增、删、改、查

1）在接口中定义增、删、改、查操作

UserDao 接口定义了增、删、改、查操作：增加一个用户、获取全部用户、根据用户 id 获取一个用户、更新一个用户、根据 id 删除一个用户。其关键代码如下：

```
public interface UserDao {
```

```java
    public void add(User user);
    public List<User> listAll();
    public User getUserById(int id);
    public void update(User user);
    public void delete(int id);
}
```

2）在 DAO 层实现类注入 JdbcTemplate

在 DAO 层实现类 UserDaoImpl 使用注解 @Autowired 注入 JdbcTemplate。UserDaoImpl 的关键代码如下：

```java
@Component
public class UserDaoImpl implements UserDao {
    @Autowired
    private JdbcTemplate jdbcTemplate;
    …
}
```

3）在 DAO 层实现类使用 JdbcTemplate 完成增、删、改、查

UserDao 接口的实现类 UserDaoServiceImpl 使用 JdbcTemplate 完成增、删、改、查操作，其关键代码如下：

```java
@Override
    public void add(User user) {
        String sql = "insert into user(name) values(?)";
        Object[] args = new Object[] { user.getName() };
        this.jdbcTemplate.update(sql, args);
    }
    @Override
    public List<User> listAll() {
        String sql = "select * from user";
        List<User> all = jdbcTemplate.query(sql, new UserMapper());
        return all;
    }
    @Override
    public User getUserById(int id) {
        String sql = "select * from user where id = ?";
        User user = jdbcTemplate.queryForObject(sql, new Object[] { 1 },
              new UserMapper());
        return user;
    }
    @Override
    public void update(User user) {
        String sql = "update user set name = ? where id = ?";
        jdbcTemplate.update(sql, new Object[] { user.getName(), 1 });
```

```java
    }
    @Override
    public void delete(int id) {
        String sql ="delete from user where id =?";
        jdbcTemplate.update(sql, new Object[] { id });

    }

    class UserMapper implements RowMapper<User>{
        public User mapRow(ResultSet rs, int rowNum) throws SQLException {
            User user =new User();
            user.setId(rs.getInt("id"));
            user.setName(rs.getString("name"));
            return user;
        }
    }
```

add(User user)方法表示增加一个用户。jdbcTemplate. update(sql, args)方法有两个参数：第一个是带占位符的 SQL 语句，第二个是参数数组对象。

listAll()方法表示获取全部用户。jdbcTemplate. queryForObject(sql, new Object[]{1}, new UserMapper())带有 3 个参数：第一个是带占位符的 SQL 语句，第二个是参数数组对象，第三个是 RowMapper 对象 UserMapper。

Spring 中的 RowMapper 将每一行数据封装成用户定义的类型。通过内部类实现 RowMapper 接口的 mapRow 方法实现。

getUserById(int id) 方法表示根据用户 id 获取一个用户。jdbcTemplate. queryForObject(sql, new Object[] { 1 }, new UserMapper())方法带有 3 个参数，参数描述与 listAll()方法相同。

update(User user)和 delete(int id)方法分别表示更新一个用户、根据 id 删除一个用户。jdbcTemplate. update()方法参数描述与 add(User user)方法中的 jdbcTemplate. update()方法相同。

5. 定义服务层接口和实现类

服务层接口 UserService 定义了服务层相关操作，其关键代码如下：

```java
public interface UserService {
    public void add(User user);
    public List<User> listAll();
    public User getUserById(int id);
    public void update(User user);
    public void delete(int id);
}
```

服务层实现类 UserServiceImpl 注入 DAO 层实现类对象并实现数据访问相关的增、

删、改、查操作，其关键代码如下：

```
@Component("userService")
public class UserServiceImpl implements UserService {
    private UserDao userDao;
    @Autowired
    public UserServiceImpl(UserDao userDao) {
        this.userDao = userDao;
    }
    @Override
    public void add(User user) {
        userDao.add(user);
    }
    @Override
    public List<User> listAll() {
        return userDao.listAll();
    }
    @Override
    public User getUserById(int id) {
        User user = userDao.getUserById(id);
        return user;
    }
    @Override
    public void update(User user) {
        userDao.update(user);
    }
    @Override
    public void delete(int id) {
        userDao.delete(id);
    }
}
```

6. 测试 JdbcTemplate

编写测试类 JdbcTemplateTest，其关键代码如下：

```
public class JdbcTemplateTest {
    public static void main(String[] args) {
        ApplicationContext context =
            new ClassPathXmlApplicationContext("jdbctemplate/beans.xml");
        UserService userService = (UserService) context.getBean("userService");
        // 1. 增加一条记录
        userService.add(new User(1, "aop"));
        // 2. 根据用户 id 获得用户对象 user
```

```
        User user =userService.getUserById(1);
        System.out.println(user.getId() +"-" +user.getName());

        // 3.获得全部用户
        List<User>all =userService.listAll();
        for (User u : all) {
            System.out.println(u.getId() +"-" +u.getName());
        }

        // 4.根据用户 id 获得用户对象 user 后,更改用户名,再更新数据库记录
        user =userService.getUserById(1);
        System.out.println(user.getId() +"-" +user.getName());
        user.setName("Mike");
        userService.update(user);
        user =userService.getUserById(1);
        System.out.println(user.getId() +"-" +user.getName());

        // 根据用户 id 删除一个用户
        userService.delete(1);
    }

}
```

12.1.2 JdbcTemplate 调用存储过程

JdbcTemplate 提供了可以调用存储过程的方法。对于无输入参数和无输出参数且无返回值的简单存储过程,可以使用 jdbcTemplate.execute("{call 简单存储过程名}")进行调用。对于带输入和输出参数及返回值的存储过程,可以选择使用如下两个方法之一进行调用。

```
execute(String callString, CallableStatementCallback action)
```

第一个参数 callString 表示存储过程 SQL 语句;第二个参数 action 表示一个回调接口,该接口只有一个方法 doInCallStatement(CallableStatement cs),用于进行输入参数绑定、输出参数注册及返回数据处理等操作。

```
execute(CallableStatementCreator csc, CallableStatementCallback action)
```

第一个参数 csc 可以创建 CallableStatement。CallableStatementCreator 定义了一个方法 CallableStatement createCallableStatement(Connection conn),使用 Connection 实例创建 CallableStatement 对象。CallableStatementCreator 负责创建 CallableStatement 实例、绑定输入参数、注册输出参数。第二个参数用于处理存储过程的返回结果。Spring 提供工厂类 CallableStatementCreatorFactory 来创建 CallableStatement Creator 实例。

下面介绍使用 JdbcTemplate 调用无输入和无输出参数且无返回值的存储过程。

1. 创建存储过程

在 MySQL 5 中创建存储过程 addUserProc1。该存储过程实现了向 user 表插入两条记录,每条记录的 name 字段值分别为 0、1。存储过程 addUserProc1 代码如下:

```
CREATE DEFINER=`root`@`localhost` PROCEDURE `addUserProc1`()
BEGIN
    DECLARE i INT;
    SET i =0;
    WHILE i <2 DO
        INSERT INTO `user`(name) VALUES(i);
        SET i =i +1;
    END WHILE;
END
```

2. 定义 DAO 层接口和实现类

在 UserDao 接口定义中增加如下 addUsersByProc1 方法:

```
public interface UserDao {
    …
    public void addUsersByProc1();
}
```

在 UserDao 接口的实现类 UserDaoImpl 中实现方法如下:

```
@Component("userDao")
public class UserDaoImpl implements UserDao {
    @Autowired
    private JdbcTemplate jdbcTemplate;
    …
    // 调用无输入参数、无返回值的存储过程
    public void addUsersByProc1() {
        jdbcTemplate.execute("{call addUserProc1}");
    }
```

3. 测试 JdbcTemplate 调用存储过程

jdbcTemplate.execute("{call addUserProc1}")方法调用数据库存储过程 addUserProc1。测试类代码 JdbcTemplateTest2 如下:

```
public class JdbcTemplateTest2 {
    public static void main(String[] args) {
        ApplicationContext context =
```

```
            new ClassPathXmlApplicationContext("jdbctemplate/beans.xml");
        UserDao userDao = (UserDao) context.getBean("userDao");
        userDao.addUsersByProc1();
    }
}
```

12.2 事务管理

12.2.1 Spring 事务管理简介

Spring 为事务管理提供了一致的编程模板,在高层次建立了统一的事务抽象。因此无论是使用 Spring JDBC 还是 JPA 或 Hibernate 等,Spring 均可使用统一的编程模型进行事务管理。

Spring 框架中涉及事务管理的 API 大约有 100 个左右,其中最重要的是位于 org.springframework.transaction 包中的 3 个接口:TransactionDefinition、PlatformTransactionManager、TransactionStatus。

1. PlatformTransactionManager

该接口用于执行具体的事务操作。它定义的 3 个方法说明见表 12-1。

表 12-1 PlatformTransactionManager 定义的 3 个方法说明

方　　法	方法说明
TransactionStatus getTransaction (TransactionDefinition definition)	根据事务定义信息从事务环境返回一个已经存在的事务,或者创建一个新的事务,并用 TransactionStatus 描述这个事务的状态
void commit(TransactionStatus status)	根据事务的状态提交事务。如果事务状态已经被标识为 rollback-only,则该方法将执行一个回滚事务的操作
void rollback(TransactionStatus status)	回滚事务。当 commit()方法抛出异常时,rollback()方法会被隐式调用

根据底层所使用的不同的持久化 API 或框架,PlatformTransactionManager 的主要实现类见表 12-2。

表 12-2 PlatformTransactionManager 的主要实现类

事务管理器	说　　明
DataSourceTransactionManager	适用于使用 JDBC 和 MyBatis 进行数据持久化操作
HibernateTransactionManager	适用于使用 Hibernate 进行数据持久化操作
JpaTransactionManager	适用于使用 JPA 进行数据持久化操作
JtaTransactionManager	适用于分布式事务或者没有其他的事务管理器的场景

续表

事务管理器	说　明
JdoTransactionManager	适用于 JDO 进行持久化操作
JmsTransactionManager	适用于 JMS 事务管理

2. TransactionDefinition

该接口描述事务的隔离级别、超时时间、是否为只读事务、事务传播规则等控制事务具体行为的事务属性。Spring 提供了一个默认的实现类 DefaultTransactionDefinition，该类适用于大多数情况。如果该类不能满足需求，则可以通过实现 TransactionDefinition 接口来实现自己的事务定义。TransactionDefinition 接口定义的 5 个方法说明见表 12-3。

表 12-3　TransactionDefinition 接口定义的 5 个方法说明

方　法	方法说明
int getPropagationBehavior()	返回事务传播行为
int getIsolationLevel()	返回事务隔离级别
String getName()	返回该事务的名称
int getTimeout()	返回单位是秒的事务超时时间
boolean isReadOnly()	返回事务是否是只读的

事务隔离级别是指若干个并发的事务之间的隔离程度。TransactionDefinition 接口定义的 5 个表示隔离级别的常量见表 12-4。

表 12-4　TransactionDefinition 接口定义的 5 个表示隔离级别的常量

事务隔离级别	说　明
ISOLATION_DEFAULT	默认隔离级别。表示使用底层数据库的默认隔离级别。对大部分数据库默认隔离级别为 ISOLATION_READ_COMMITTED
ISOLATION_READ_UNCOMMITTED	读未提交。表示一个事务可以读取另一个事务修改但尚未提交的数据；不能防止脏读和不可重复读。较少使用
ISOLATION_READ_COMMITTED	读已提交。表示一个事务只能读取另一个事务已经提交的数据。可以防止脏读。经常使用
ISOLATION_REPEATABLE_READ	可重复读。表示一个事务可以多次重复执行某个查询，并且每次返回的记录都相同。即使在多次查询之间有新增的数据满足该查询，这些新增的记录也会被忽略。可以防止脏读和不可重复读。较少使用
ISOLATION_SERIALIZABLE	串行化。所有的事务依次逐个执行，这样事务之间就完全不可能产生干扰。该级别可以防止脏读、不可重复读以及幻读，但是这将严重影响程序的性能。较少使用

事务传播行为是指如果在开始当前事务之前，一个事务上下文已经存在，那么此时有

若干选项可以指定一个事务性方法的执行行为。TransactionDefinition 定义的 7 个表示传播行为的常量说明见表 12-5。

表 12-5 TransactionDefinition 接口定义的 7 个表示传播行为的常量说明

事务传播行为	说　明
PROPAGATION_REQUIRED	如果当前存在事务，则加入该事务；如果当前没有事务，则创建一个新的事务
PROPAGATION_REQUIRES_NEW	创建一个新的事务，如果当前存在事务，则把当前事务挂起
PROPAGATION_SUPPORTS	如果当前存在事务，则加入该事务；如果当前没有事务，则以非事务的方式继续运行
PROPAGATION_NOT_SUPPORTED	以非事务方式运行，如果当前存在事务，则把当前事务挂起
PROPAGATION_NEVER	以非事务方式运行，如果当前存在事务，则抛出异常
PROPAGATION_MANDATORY	如果当前存在事务，则加入该事务；如果当前没有事务，则抛出异常
PROPAGATION_NESTED	如果当前存在事务，则创建一个事务作为当前事务的嵌套事务来运行；如果当前没有事务，则该取值等价于 TransactionDefinition. PROPAGATION_REQUIRED

事务超时就是指一个事务所允许执行的最长时间。如果超过该时间限制但事务还没有完成，则自动回滚事务。在 TransactionDefinition 中以 int 的值来表示超时时间，其单位是秒。

事务的只读属性是指对事务性资源进行只读操作或者是读写操作。所谓事务性资源，就是指那些被事务管理的资源，例如，数据源、JMS 资源、自定义的事务性资源等。如果确定只对事务性资源进行只读操作，那么可以将事务标志为只读的，以提高事务处理的性能。在 TransactionDefinition 中以 boolean 类型来表示该事务是否只读。

3. TransactionStatus

TransactionStatus 表示一个事务的运行状态。事务管理器可以通过该接口获取事务运行期的状态信息，也可以通过该接口间接地回滚事务。它相比于在抛出异常时回滚事务的方式更具有可控性。TransactionStatus 的主要方法说明见表 12-6。

表 12-6 TransactionStatus 的主要方法说明

方　法	说　明
boolean isNewTransaction()	判断当前事务是否是一个新的事务，如果返回 false，则表示当前事务是一个已经存在的事务，或当前操作未运行在事务环境中
void setRollbackOnly()	设置当前事务为 rollback-only。通过该标识通知事务管理器只能将事务回滚，事务管理器将通过显式调用回滚命令或抛出异常的方式回滚事务
boolean isRollbackOnly()	判断当前事务是否被标识为 rollback-only

Spring 支持编程式事务管理和声明式事务管理。编程式事务管理是使用代码进行事务管理。声明式事务管理是使用注解或 XML 配置来管理事务。

实践中尽可能声明式事务管理。尽管不如编程式事务管理灵活，但声明式事务管理允许通过代码控制事务。Spring 支持使用 Spring AOP 框架的声明式事务管理。

12.2.2 编程式事务

Spring 提供了两种方式进行编程式事务管理：使用 TransactionTemplate；直接使用一个 PlatformTransactionManager 实现。Spring 建议使用第一种方式。第二种方式类似于 JTA 的 UserTransaction API，在此不做介绍。

下面介绍使用 TransactionTemplate 进行事务管理。在 UserServiceImpl 实现类的 addTwoUsers(final List<User> users)方法负责向数据库 user 表插入两条记录。插入两条记录作为一个事务进行，要么一起插入两条记录，要么一条也不会插入。UserServiceImpl 类的关键代码如下：

```
@Component("userService")
public class UserServiceImpl implements UserService {
    private UserDao userDao;
    private TransactionTemplate transactionTemplate;

    @Autowired
    public UserServiceImpl(UserDao userDao) {
        this.userDao =userDao;
    }

    @Autowired
    public void setTransactionTemplate(TransactionTemplate transactionTemplate) {
        this.transactionTemplate =transactionTemplate;
    }

    //一次添加两个用户的事务操作
    public void addTwoUsers(final List<User>users) {
        transactionTemplate.execute(new TransactionCallbackWithoutResult() {
            @Override
            public void doInTransactionWithoutResult(TransactionStatus status) {
                add(users.get(0));
                //int i =5 / 0;
                add(users.get(1));
            }
        });
    }
    ...
```

}

UserServiceImpl 中使用 @ Autowired 注解注入了 TransactionTemplate transactionTemplated 和 UserDao userDao。

TransactionTemplate. execute (TransactionCallback action) 的回调接口 TransactionCallback 中定义需要以事务方式进行的业务逻辑。

TransactionCallback 接口只有一个方法：doTransaction (TransactionStatus status)。

如果事务操作不需要返回结果，需要使用 TransactionCallback 的子接口 TransactionCallbackWithoutResult，实现其 public void doInTransactionWithoutResult (TransactionStatus status)方法。

如果事务操作需要返回结果，则需要使用 TransactionCallback 接口实现其 public Object doInTransaction(TransactionStatus status)方法。

本案例不需要返回结果，因而使用的是 TransactionCallbackWithoutResult 子接口及其 doInTransactionWithoutResult()方法。

在两次添加用户记录之间，int i = 5 / 0 会抛出异常，用于进行事务测试。在该测试代码被注解后，程序正常执行会向数据库 user 表以事务方式一次增加两条记录；取消该测试代码注解，则抛出异常，以事务方式进行的增加两条记录的操作不会插入任何记录。

测试类 JdbcTemplateTest 的关键代码如下：

```
public class JdbcTemplateTest {
    public static void main(String[] args) {
        ApplicationContext context =
            new ClassPathXmlApplicationContext("beans.xml");
        UserService userService = (UserService) context.getBean("userService");
        // 1. 增加两条记录
        User user1 =new User();
        user1.setName("user-1");
        User user2 =new User();
        user2.setName("user-2");
        List<User>twoUsers =new ArrayList<User>();
        twoUsers.add(user1);
        twoUsers.add(user2);
        userService.addTwoUsers(twoUsers);
    }
}
```

Spring 配置文件 beans. xml 的关键代码如下：

```
<?xml version="1.0" encoding="UTF-8"?>
<beans …
  <context:component-scan base-package="dao,service" />
  <!--配置数据源 -->
```

```
<bean id="dataSource"
   class="org.springframework.jdbc.datasource.DriverManagerDataSource">
   <property name="driverClassName" value="com.mysql.jdbc.Driver"/>
   <property name="url" value="jdbc:mysql://localhost:3306/testdb"/>
   <property name="username" value="root"/>
   <property name="password" value="root"/>
</bean>
<!--配置 JDBC 事务管理器 -->
<bean id="transactionManager"
   class="org.springframework.jdbc.datasource.DataSourceTransactionManager">
   <property name="dataSource"  ref="dataSource" />
</bean>
<!--配置事务模板 TransactionTemplate -->
<bean id="transactionTemplate"
   class="org.springframework.transaction.support.TransactionTemplate">
   <property name="transactionManager" ref="transactionManager" />
   <!--ISOLATION_DEFAULT 表示由使用的数据库决定隔离级别  -->
   <property name="isolationLevelName" value="ISOLATION_DEFAULT"/>
   <property name="propagationBehaviorName"
     value="PROPAGATION_REQUIRED"/>
</bean>
<!--配置 JdbcTemplate -->
<bean id="jdbcTemplate" class="org.springframework.jdbc.core.JdbcTemplate">
   <property name ="dataSource" ref="dataSource" />
</bean>
</beans>
```

12.2.3　声明式事务

Spring 对声明式事务的支持是通过 Spring AOP 框架实现的。Spring 提供了两种使用声明式事务的方式：Spring 2.0 之前使用 Spring AOP 和 TransactionProxyFactoryBean 的代理 Bean 方式实现声明式事务；现在主要使用 XML 配置声明式事务或使用注解配置声明式事务。此处介绍第二种方式。

1. 使用 XML 配置声明式事务

此处还是以实现类 UserServiceImpl 的 addTwoUsers(final List<User> users)方法为例说明使用 XML 的 Spring 声明式事务。该方法向数据库 user 表插入两条记录作为一个事务，要么一起插入两条记录，要么一条也不会插入。UserServiceImpl 类的关键代码如下：

```
@Component("userService")
public class UserServiceImpl implements UserService {
```

```
    private UserDao userDao;
    @Autowired
    public UserServiceImpl(UserDao userDao) {
        this.userDao =userDao;
    }
    //一次添加两个用户的事务操作
    @Override
    public void addTwoUsers(final List<User>users) {
        add(users.get(0));
        //int i =5 / 0;
        add(users.get(1));
    }
    …
}
```

addTwoUsers()方法中调用 add(users.get(0))和 add(users.get(1))方法插入两条记录。

在配置文件 beans.xml 中引入 tx 和 aop 命名空间，如下面加粗部分的代码所示：

```
<?xml version="1.0" encoding="UTF-8"?>
<beans xmlns="http://www.springframework.org/schema/beans"
    xmlns:xsi="http://www.w3.org/2001/XMLSchema-instance"
    xmlns:context="http://www.springframework.org/schema/context"
    xmlns:aop="http://www.springframework.org/schema/aop"
    xmlns:tx="http://www.springframework.org/schema/tx"
    xsi:schemaLocation="http://www.springframework.org/schema/beans
    http://www.springframework.org/schema/beans/spring-beans-4.0.xsd
    http://www.springframework.org/schema/aop
    http://www.springframework.org/schema/aop/spring-aop-4.0.xsd
    http://www.springframework.org/schema/tx
    http://www.springframework.org/schema/tx/spring-tx-4.0.xsd
    http://www.springframework.org/schema/context
    http://www.springframework.org/schema/context/spring-context-4.0.xsd" >
```

(1) 在 beans.xml 中，首先配置开启自动扫描、配置数据源、配置 JdbcTemplate，其关键的代码如下：

```
<?xml version="1.0" encoding="UTF-8"?>
<beans …
  <!--开启自动扫描 -->
  <context:component-scan base-package="dao,service" />
  <!--配置数据源 -->
  <bean id="dataSource"
     class="org.springframework.jdbc.datasource.DriverManagerDataSource">
     <property name="driverClassName" value="com.mysql.jdbc.Driver"/>
```

```
        <property name="url" value="jdbc:mysql://localhost:3306/testdb"/>
        <property name="username" value="root"/>
        <property name="password" value="root"/>
    </bean>
    <!--配置 JDBC 事务管理器 -->
    <bean id="transactionManager"
        class="org.springframework.jdbc.datasource.DataSourceTransactionManager">
        <property name="dataSource"  ref="dataSource" />
    </bean>
    <!--配置 JdbcTemplate -->
    <bean id="jdbcTemplate" class="org.springframework.jdbc.core.JdbcTemplate">
        <property name ="dataSource" ref="dataSource" />
    </bean>
</beans>
```

(2) 接着使用<tx:advice>定义事务属性，其关键的代码如下：

```
<?xml version="1.0" encoding="UTF-8"?>
<beans …
    …
    <tx:advice id="txAdvice"  transaction-manager="transactionManager">
        <tx:attributes>
            <tx:method name="addTwoUsers"/>
        </tx:attributes>
    </tx:advice>
</beans>
```

<tx:advice>的 transaction-manager 属性指定事务管理器。

<tx:attributes>用于定义事务属性，其子标签<tx:method name="addTwoUsers"/>为 name 属性指定的 addTwoUsers 方法定义事务参数。<tx:method>定义方法的事务策略的属性说明见表 12-7。

表 12-7　<tx:method>定义方法的事务策略的属性说明

隔离级别	说　明
isolation	指定事务的隔离级别
propagation	指定事务的传播规则
read-only	指定事务为只读
回滚规则： rollback-for no-rollback-for	rollback-for 指定事务对于哪些检查异常类型应当回滚而不提交 no-rollback-for 指定事务对于哪些异常应当继续执行而不回滚
timeout	指定事务超时时间

(3) 最后使用<aop:config>定义以 Spring AOP 增强实施事务管理。

```
<?xml version="1.0" encoding="UTF-8"?>
<beans …
    …
    <aop:config>
        <aop:pointcut id="addTwoUsers"
          expression="execution(* service.UserServiceImpl.addTwoUsers(..))"/>
        <aop:advisor advice-ref="txAdvice" pointcut-ref="addTwoUsers"/>
    </aop:config>
</beans>
```

<aop:config>的<aop:pointcut>定义了执行 service.UserServiceImpl 类的 addTwoUsers()方法时实施事务增强。

测试类 JdbcTemplateTest 代码如下：

```
public class JdbcTemplateTest {
    public static void main(String[] args) {
        ApplicationContext context =
            new ClassPathXmlApplicationContext("beans.xml");
        UserService userService = (UserService) context.getBean("userService");
        // 1. 增加两条记录
        User user1 =new User();
        user1.setName("user-1");
        User user2 =new User();
        user2.setName("user-2");
        List<User>twoUsers =new ArrayList<User>();
        twoUsers.add(user1);
        twoUsers.add(user2);
        userService.addTwoUsers(twoUsers);
    }
}
```

2. 使用注解配置声明式事务

在此还是以 UserServiceImpl 实现类的 addTwoUsers()方法向数据库 user 表插入两条记录作为一个事务为例，说明使用注解配置 Spring 声明式事务，其关键代码如下：

```
@Component("userService")
public class UserServiceImpl implements UserService {
    private UserDao userDao;
    @Autowired
    public UserServiceImpl(UserDao userDao) {
        this.userDao =userDao;
    }
    // 一次添加两个用户的事务操作
    @Override
```

```
    @Transactional(propagation =Propagation.REQUIRED, readOnly =false)
    public void addTwoUsers(final List<User>users) {
        add(users.get(0));
        // int i =5 / 0;
        add(users.get(1));
    }
    …
}
```

@Transactional(propagation = Propagation. REQUIRED, readOnly = false)注解用于 addTwoUsers()方法之前,表示对该方法实施 Spring AOP 事务增强。@Transactional 注解也可以用于类前,表示对类中的所有方法实施 Spring AOP 事务增强。

@Transactional 注解只是提供元数据,本身不能完成事务切面织入。因此还需要在 Spring 配置文件中进行配置,其关键代码如下:

```
<?xml version="1.0" encoding="UTF-8"?>
<beans …
    <context:component-scan base-package="service" />
    <!--开启事务的注解支持 -->
    <tx:annotation-driven transaction-manager="transactionManager"/>
    <!--配置数据源 -->
    <bean id="dataSource"
        …
    </bean>
    <!--定义事务管理器(声明式的事务) -->
    <bean id="transactionManager"
        …
    </bean>
    <!--配置 JdbcTemplate -->
    <bean id="jdbcTemplate" class="org.springframework.jdbc.core.JdbcTemplate">
        <property name ="dataSource" ref="dataSource" />
    </bean>
</beans>
```

与基于 XML 使用 Spring 声明式事务的不同之处是,在配置文件中删除了<tx:advice>和<aop:config>元素。配置文件使用<tx:annotation-driven>元素开启 Spring 事务的注解支持,这样 Spring 会自动扫描所有使用@Transactional 注解的 Bean,为该 Bean 的所有方法或某个方法实施事务增强。

12.3 Spring 整合 JPA

基于 JPA 的应用程序需要使用 EntityManagerFactory 的实现类来获得 EntityManager 对象,再使用 EntityManager 对象对实体对象进行操作。

JPA 定义了两种类型的实体管理器：应用管理的实体管理器和容器管理的实体管理器。应用管理的实体管理器，需要应用程序使用 EntityManagerFactory 来创建 EntityManager、使用 EntityManager 手动管理事务并负责 EntityManager 的关闭。容器管理的实体管理器，应用程序不与 EntityManagerFactory 交互。在容器中配置 EntityManagerFactory，通过 JNDI 或者依赖注入来获取 EntityManager。

Spring 中使用 JPA 的关键是在 Spring 应用上下文中将 EntityManagerFactory 配置为一个 Bean。Spring 提供 3 种方式整合 JPA。

12.3.1 配置 LocalEntityManagerFactoryBean

用于获取应用管理的实体管理器。Spring 配置文件 beans.xml 配置如下：

```
<bean id="entityManagerFactory"
   class="org.springframework.orm.jpa.LocalEntityManagerFactoryBean">
   <property name="persistenceUnitName" value="pu"/>
</bean>
```

LocalEntityManagerFactoryBean 根据 JPA PersistenceProvider 自动检测配置文件进行工作，一般从"META-INF/persistence.xml"读取配置信息。

这种方式最简单，但是不能设置 Spring 中定义的 DataSource，且不支持 Spring 管理的全局事务，适用于那些仅使用 JPA 进行数据访问的项目。实际上，该方式常常用于独立的应用程序和测试环境。

12.3.2 配置从 JNDI 获取 EntityManagerFactory

用于 Spring 应用程序部署在应用服务器(如 Tomcat)上，并且使用 JNDI 创建 EntityManagerFactory 之后，这时使用 Spring jee 命名空间下的<jee:jndi-lookup>元素来获取对 EntityManagerFactory 的引用。这种方式在 Spring 事务管理时一般要使用 JTA 事务管理。Spring 配置文件 beans.xml 配置如下：

```
<beans xmlns="http://www.springframework.org/schema/beans"
   xmlns:xsi="http://www.w3.org/2001/XMLSchema-instance"
   xmlns:jee="http://www.springframework.org/schema/jee"
   xsi:schemaLocation="
   http://www.springframework.org/schema/beans
   http://www.springframework.org/schema/beans/spring-beans-3.0.xsd
   http://www.springframework.org/schema/jee
   http://www.springframework.org/schema/jee/spring-jee-3.0.xsd">
   <jee:jndi-lookup id="entityManagerFactory" jndi-name="persistence/
   persistenceUnit"/>
</beans>
```

12.3.3 配置 LocalContainerEntityManagerFactoryBean

用于获取 Spring 容器管理的实体管理器。此处以 Eclipse 的 Java 项目 ch12-03-03 为例说明 Spring 容器管理的实体管理器的使用。

1. 创建 Eclipse 的 Java 项目

创建 Eclipse 的 Java 项目 ch12-03-03。

配置项目构建路径添加 34 个相关 jar 包如下：

spring-framework-4.1.6.RELEASE\libs 下带 RELEASE 的 20 个 jar 文件、MySQL 下 mysql-connector-java-5.0.6.jar、common-loggin-1.2.jar、aopalliance-1.0.jar、hibernate-release-4.3.10.Final\lib\required 下的 10 个 jar、hibernate-release-4.3.10.Final\lib\jpa 下的 hibernate-entitymanager-4.3.10.Final.jar。

2. 配置数据源

在项目 ch12-03-03 的 src 下创建 Spring 配置文件 beans.xml，配置数据源如下：

```
<!--数据源配置 -->
<bean id="dataSource"
   class="org.springframework.jdbc.datasource.DriverManagerDataSource" >
   <property name="driverClassName" value="${db.driverClass}"/>
   <property name="url" value="${db.jdbcUrl}" />
   <property name="username" value="${db.user}" />
   <property name="password" value="${db.password}" />
</bean>
<bean id="propertyPlaceholderConfigurer"
   class="org.springframework.beans.factory.config.PropertyPlaceholderConfigurer">
   <property name="locations">
      <list>
         <value>classpath:db.properties</value>
      </list>
   </property>
</bean>
```

通过配置<bean id="propertyPlaceholderConfigurer">实现读取 classpath 下的 db.properties 文件中定义的数据源信息。PropertyPlaceholderConfigurer 可以将上下文（配置文件）中的属性值放在另一个单独的标准 Properties 文件中，在 XML 文件中用 ${key}替换指定的 properties 文件中的值。

db.properties 内容如下：

```
db.driverClass=com.mysql.jdbc.Driver
db.jdbcUrl=jdbc:mysql://localhost:3306/testDB
```

```
db.user=root
db.password=root
```

3. 配置 LocalContainerEntityManagerFactoryBean

```
<!--配置 LocalContainerEntityManagerFactoryBean-->
<bean id="entityManagerFactory"
  class="org.springframework.orm.jpa.LocalContainerEntityManagerFactoryBean">
  <!--指定数据源 -->
  <property name="dataSource" ref="dataSource"/>
  <!--指定 JPA 实现厂商-->
  <property name="jpaVendorAdapter">
    <bean id="hibernateJpaVendorAdapter"
        class="org.springframework.orm.jpa.vendor.HibernateJpaVendorAdapter">
        <property name="generateDdl" value="true" />
        <property name="database" value="MYSQL" />
        <property name="databasePlatform"
          value="org.hibernate.dialect.MySQL5InnoDBDialect" />
        <property name="showSql" value="true" />
    </bean>
  </property>
  <!--指定 Entity 实体类包路径:扫描 entity 包中带@Entity 注解的类 -->
  <property name="packagesToScan" >
    <array>
        <value>entity</value>
    </array>
  </property>
</bean>
```

4. 配置 JPA 事务管理器

```
<!--JPA 事务管理器  -->
<bean id="transactionManager"
  class="org.springframework.orm.jpa.JpaTransactionManager">
  <property name="entityManagerFactory" ref="entityManagerFactory"/>
</bean>
```

5. 开启自动扫描和注解事务

```
<!--开启自动扫描:自动扫描 dao 包中带@Component 等注解的类并为其创建 Bean -->
<context:component-scan base-package="dao" />
<!--开启注解事务管理:为带@Transactional 注解的 Bean 自动配置为声明式事务支持-->
<tx:annotation-driven transaction-manager="transactionManager"
  proxy-target-class="true" />
```

proxy-target-class 属性用于指定 Spring 采用的是基于接口的代理还是基于类的代理。如果 proxy-target-class 属性值被设置为 true，那么 Spring 采用的是基于类的代理（这时需要 cglib 库）；如果 proxy-target-class 属性值被设置为 false 或者这个属性被省略，那么 Spring 采用的是 JDK 的基于接口的代理。

6. 编写实体类 User、DAO 层接口 UserDao 和实现类 UserDaoImpl

实体类 User 的关键代码如下：

```
package entity;
…
@Entity
public class User implements Serializable {
    @Id
    @GeneratedValue
    private int id;
    @Column(name ="name", nullable =false, length =50,
        insertable =true, updatable =true)
    private String name;
    private String password;
    …
}
```

UseDao 接口的关键代码如下：

```
package dao;
…
public interface UserDao {
    public User save(User user);
    public void remove(int id);
    public User update(User user);
    public User get(int id);
    public List<User> findAll();
}
```

Dao 层实现类 UserDaoImpl 的关键代码如下：

```
package dao;
…
@Repository
@Transactional
public class UserDaoImpl implements UserDao {
    @PersistenceContext
    private EntityManager em;
    @Override
    public User save(User user) {
```

```
        em.persist(user);
        return user;
    }
    ...
}
```

其中以下代码：

```
@PersistenceContext
private EntityManager em;
```

表示容器采用依赖注入方式注入了 EntityManager 对象 em。

7. 编写测试类

测试类 UserDaoTest 的关键代码如下：

```
...
public class UserDaoTest {
    public static void main(String[] args) {
        add();
    }
    public static void add() {
        ApplicationContext context =
            new ClassPathXmlApplicationContext("beans.xml");
        UserDao userDao = (UserDao) context.getBean("userDaoImpl");
        User user =new User();
        user.setName("Alice");
        user.setPassword("Alice123");
        userDao.save(user);
    }
}
```

12.3.4 Spring 整合 JPA 时使用 Spring Data

Spring Data 是 Spring 的一个子项目，其目标是统一和简化各种持久化编程，实现无论是何种持久化存储，数据访问对象都会提供对单一域对象的增、删、改、查和分页处理等基本操作。Spring Data 提供了各种持久化存储的统一接口（如 CrudRepository、PagingAndSortingRepository）以及各种持久化存储的实现。

Spring Data 包含多个子项目，常见的几个子项目说明如下：

- Commons——提供 Spring Data 项目的核心基础。
- Key-Value——提供对基于 Map 存储支持。
- Neo4j——提供对基于对象图像存储 Neo4j 的支持。
- JDBC——提供对基于 JDBC 的数据访问简化编程支持。

• JPA——提供对基于 JPA 数据访问简化编程的支持。
• Hadoop——提供基于 Hadoop 的作业配置和 POJO 编程模型的 MapReduce 作业的支持。

下面以 Eclipse 的 Java 项目 ch12-03-04 为例说明 Spring 整合 JPA 时使用 Spring Data 的一般步骤和 Spring Data 相关技术知识。

1. 创建 Eclipse 的 Java 项目

创建 Eclipse 的 Java 项目 ch12-03-04。

配置项目构建路径添加相关 jar 包。添加的相关 jar 包除了 12.3.3 节 ch13-03-03 项目的 34 个 jar 文件,还包括 software\spring-data 下的两个 jar 文件、log4j-1.2.17.jar、slf4j-api-1.5.8.jar 和 slf4j-log4j12-1.5.8.jar。

2. 编写配置文件

在项目的 src 下创建和编写配置文件 beans.xml 和 db.properties。db.properties 内容与 12.3.3 节的 ch13-03-03 项目相同。beans.xm 增加了如下一行:

```
<jpa:repositories base-package="dao"/>
```

<jpa:repositories>会扫描它的 dao 包来查找扩展自 Spring Data JPA Repository 接口的所有接口。如果发现了扩展自 Repository 的接口,在应用启动的时候就会自动生成这个接口的实现。此处会生成 UserDao 接口的实现类。

3. 编写实体类

实体类 User 代码与 12.3.3 节 ch13-03-03 项目中的 User 代码相同。

4. 编写 UserDao 接口

UserDao 接口代码如下:

```
package dao;
import org.springframework.data.jpa.repository.JpaRepository;
import entity.User;
public interface UserDao extends JpaRepository<User, Long>{

}
```

Spring Data JPA 中 Repository 相关接口的继承层次结构如图 12-1 所示。

JpaRepository 从父接口继承的可用方法有几十个,其常用方法说明见表 12-8。

通过扩展接口 JpaRepository<User, Long>,UserDao 接口具备了对 User 对象增、删、改、查相关的几十个方法操作。Spring Data JPA 会在应用启动的时候完成 UserDao 接口实现类的创建和对象的创建,因此不需要编写 UserDao 的实现类代码。

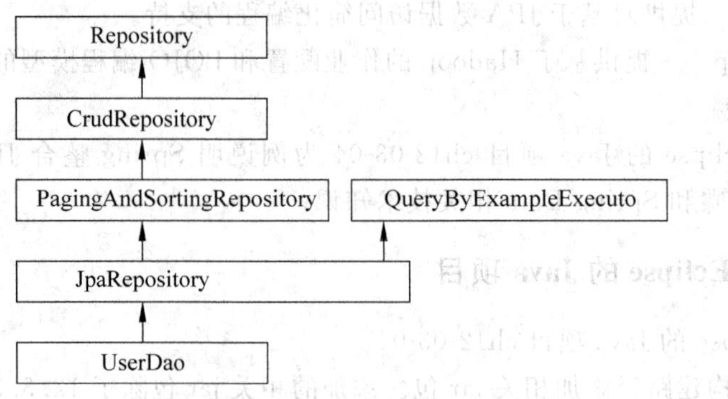

图 12-1 Repository 相关接口的继承层次结构

表 12-8 JpaRepository 从父接口继承的常用方法说明

修饰符、返回值类型	方法名、参数	说明
void	deleteAllInBatch()	使用一个批处理操作删除所有实体
List<T>	findAll()	查找全部实体对象
<S extends T> List<S>	findAll(Example<S> example)	查找全部实体对象
<S extends T> long	count(Example<S> example)	返回符合条件的对象个数
<S extends T>boolean	exists(Example<S> example)	判断是否有符合条件的对象
<S extends T> Page<S>	findAll(Example<S> example, Pageable pageable)	查找一批对象且排序和分页
<S extends T> S	findOne(Example<S> example)	查找一个对象
<S extends T> S	save(S entity)	保存或更新一个实体对象。如果实体对象没有 ID,则执行保存操作,否则执行更新操作
Optional<T>	findById(ID primaryKey)	根据 ID 查找一个实体对象
void	delete(T entity)	删除参数指定的实体对象
boolean	existsById(ID primaryKey)	判断 ID 指定的实体对象是否存在

5. 编写 UserServiceManager 类

通常在针对数据库进行主要操作的应用中,为了简化编程,服务层可以省略接口,仅仅提供一个实现类。此处服务层没有编写接口 UserService,而仅仅提供了实现类 UserServiceManager,其关键代码如下:

```
package service;
...
@Service("userService")
public class UserServiceManager {
    //推荐用 Resource 来替代 AutoWrite 注解
```

```
    @Resource
    private UserDao userDao;
    //新增用户
    public void save(User user) {
        userDao.save(user);
    }
    //删除用户,参数也可以为一个含有 id 的 User 对象
    public void delete(Long id) {
        userDao.delete(id);
    }
    //查询所有 user 对象,findOne 为查询单个对象
    public List<User> findAll() {
        return (List<User>) userDao.findAll();
    }
    public Page<User> findAllByPage(PageRequest page) {
        return (Page<User>) userDao.findAll(page);
    }
}
```

6. 编写测试类

测试类 UserServiceTest 的关键代码如下：

```
public class UserServiceTest {
    public static void main(String[] args) {
        add();
    }
    public static void add() {
        ApplicationContext context =
            new ClassPathXmlApplicationContext("beans.xml");
        UserServiceManager userService =
            (UserServiceManager) context.getBean("userService");
        User user = new User();
        user.setName("Alice");
        user.setPassword("Alice123");
        userService.save(user);
    }
}
```

12.3.5 Spring Data JPA 的自定义查询

有时候继承 JpaRepository 接口的方法在查询时不能满足要求，需要自定义查询。Spring Data JPA 中自定义查询有 3 种方式。

1. 自定义符合 Spring Data JPA 命名规范的查询方法

直接在接口中自定义符合 Spring Data JPA 命名规范的查询方法,不用编写实现代码。

例如,需要在 UserDao 接口增加自定义查询方法 User findByName(String name),该方法根据名字查询满足条件的实体对象。该方法定义如下:

```
public interface UserDao extends JpaRepository<User, Long>  {
    User findByName(Stringname);
}
```

不需要编写 User findByName(String name)方法的实现,Spring Data JPA 负责创建这个方法的实现。

自定义符合 Spring Data JPA 命名规范的查询方法时,查询方法的命名规则如下:

```
查询动词 [主题] By 限定短语
```

(1) 查询动词。

包括 get、read、find 和 count。get、read 和 find 这 3 个动词含义相同,用于表示查询数据并返回对象。动词 count 会返回匹配对象的数量。

(2) 主题。

主题可选的,通常是实体类类名,这种情况下可以省略。如果主题以 DISTINCT 开头,则表示查询不重复的记录。

(3) 限定短语。

限定短语格式:

```
属性名操作符[属性名][操作符] …
```

常用的操作符及其使用说明见表 12-9。

表 12-9 常用的操作符及其使用说明

操作符	方法命名的例子	对应的 JPA 查询语言片段
And	findByLastnameAndFirstname(String lastname, String firstname)	…where x. lastname=? 1 and x. firstname=? 2
Or	findByLastnameOrFirstname(String lastname, String firstname)	…where x. lastname= ? 1 or firstname= ? 2
Between	findByBirthdayBetween(Datebegin, Date end)	…where x. birthday between ? 1 and ? 2
LessThan	findByIdLessThan(Long id)	…where x. id<? 1
GreaterThan	findByIdGreaterThan(Long id)	…where x. id>? 1
After	findByBirthdayAfter(Date date)	…where x. birthday>? 1

续表

操作符	方法命名的例子	对应的 JPA 查询语言片段
Before	findByBirthdayBefore(Date date)	…where x. birthday<? 1
IsNull	findByBirthdayIsNull()	…where x. birthday is null
IsNotNull	findByBirthdayIsNotNull()	…where x. birthday not null
Like	findByNameLike(String name)	…where x. name like ? 1
NotLike	findByNameNotLike(String name)	…where x. name not like ? 1
StaringWith	findByNameStaringWith(String name)	…where x. name like ? 1 (%参数值)
EndingWith	findByNameEndingWith(String name)	…where x. name like ? 1 (参数值%)
Containing	findByNameContaining(String name)	…where x. name like ? 1 (%参数值%)
OrderBy	findByNameOrderByBirthdayDesc(String name)	…where x. name=? 1 order by x. birthday desc
Not	findByNameNot(String name)	…where x. name<>? 1
In	findByIdIn(Collection<Long> ids)	…where x. id in ? 1
NotIn	findByIdNotIn(Collection<Long> ids)	…where x. id not in ? 1
True	findByStateTrue()	…where x. state=true
False	findByStateFalse ()	…where x. state=false

2. 在自定义查询方法上使用@Query

可以在自定义查询方法上使用注解@Query 来指定该方法要执行的查询语句。例如，如下的 findByName()方法上使用注解@Query 来查询参数为 String name 的 User 对象。

```
@Query("select o from User o where o.name=?1")
public List<User> findByName(String name);
```

另外还可以使用注解@Query 来指定本地查询。这时需要设置@Query 的 nativeQuery 属性为 true。

```
@Query(value="select * from user where name like %?1",nativeQuery=true)
public List<User> findByName(String name);
```

注意当前版本的本地查询不支持分页和动态的排序。

3. 在自定义查询方法上使用@NamedQuery

在自定义查询方法上使用 JPA 的@NamedQuery 的一般步骤如下：

（1）在实体类上使用@NamedQuery。

```
@Entity
@NamedQuery(name ="User.findByName",
    query ="select o from User o where o.name =?1")
public class User implements Serializable {
    …
}
```

（2）在 Dao 层实现 Repository 的接口 UserDao 中定义一个同名的方法。

```
public interface UserDao extends JpaRepository<User, Long>{
    public List<User>findByName(String name);
}
```

（3）调用 findByName(String name)方法。

这时 Spring 首先查找是否有同名的命名查询 User. findByName。如果有就不会按照自定义符合 Spring Data JPA 命名规范的查询方法来解析，而是直接使用该命名查询。

12.3.6 自定义查询方法的使用顺序

Spring Data JPA 使用<jpa:repositories> 的 query-lookup-strategy 属性来指定查找顺序。query-lookup-strategy 属性有如下 3 个取值。

1. create-if-not-found

如果方法前使用@Query 指定了查询语句，则使用该语句实现查询；如果没有，则查找方法前是否使用@NamedQuery 指定的命名查询，如果找到，则使用该命名查询；如果两者都没有找到，则通过解析方法名字来创建查询。query-lookup-strategy 属性的默认值是 create-if-not-found。

2. create

按照自定义符合 Spring Data JPA 命名规范的查询方法来解析方法名方式来创建查询；忽略方法前的注解@NamedQuery 和@Query 指定的查询。

3. use-declared-query

如果方法前使用@Query 指定了查询语句，则使用该语句实现查询；如果没有，则查找方法前是否使用@NamedQuery 指定的命名查询，如果找到，则使用该命名查询；如果两者都没有找到，则抛出异常。

小 结

Spring 没有自己的持久化实现。Spring 对持久化的支持主要是提供了对其他持久化技术的集成和开发简化。在 Spring 使用 JDBC 比单独使用 JDBC 要简洁和优雅。

本章主要介绍了在 Spring 中使用 JdbcTemplate 进行数据库增、删、改、查和调用存储过程，Spring 对事务管理在高层建立了统一的抽象和 Spring 事务相关的 API，以及 Spring 对编程式事务和声明式事务的支持。最后介绍了 Spring 整合 JPA 和整合 JAP 时使用 Spring Data 简化开发。

思考与习题

1. 写出 Eclipse 下创建 Java 项目使用 JdbcTemplate 的一般步骤。

2. 编写程序使用 JdbcTemplate 完成对 Employee 实体的增、删、改、查。Employee 实体类的属性有 int id、String name、int age、String sex。

3. 写出 JdbcTemplate 调用存储过程的一般步骤。

4. Spring 事务框架中最重要的接口有哪几个？这几个接口各自的作用是什么？

5. Spring 提供的两种方式的编程式事务管理分别是什么？Spring 建议使用哪种方式？

6. 向 user 表插入一条新记录(id=2)和修改 id=1 的记录的两个方法当作一个事务进行。对上述操作编写程序，要求使用 TransactionTemplate 进行事务管理并进行测试。

7. Spring 提供的两种使用声明式事务的方式分别是什么？现在主要使用的是哪种方式？

8. 仿照实现类 UserServiceImpl 的 addTwoUsers(final List<User> users)方法，并使用 XML 配置 Spring 声明式事务来编写一个新的方法 addAndUpateUsers(final List <User> users)，该方法向数据库 user 表插入 users 中的第一条记录并更新 users 中的第二条记录；插入和更新的方法作为一个事务。

9. 使用注解配置 Spring 声明式事务完成上面第 8 题。

10. JPA 定义的两种类型的实体管理器分别是什么？

11. Spring 中使用 JPA 的关键是什么？

12. Spring 提供的 3 种整合 JPA 方式分别是什么？

13. 写出基于 Eclipse 的 Java 项目配置 LocalContainerEntityManagerFactoryBean 获取 Spring 容器管理的实体管理器的一般步骤。

14. 写出基于 Eclipse 的 Java 项目中配置 Spring 整合 JPA 时采用 Spring Data 的一般步骤。

15. Spring 整合 JPA 时采用 Spring Data，为什么不需要编写 UserDao 的实现类代码？

16. Spring Data JPA 中自定义查询的 3 种方式分别是什么？

第13章

Spring MVC

Spring 除了提供对已有的 MVC 框架如 Struts2 等的支持之外，还有自己的 MVC 框架 Spring MVC。Spring MVC 是基于 Model 2 实现的 MVC 框架。Spring MVC 框架的核心是 DispatcherServlet。DispatcherServlet 负责拦截请求，并将其分派给相应的控制器来处理。

Spring MVC 处理请求的过程简述如下。

(1) 客户端发送一个 HTTP 请求到 Web 服务器；

(2) DispatcherServlet 根据 URL 或请求参数，通过查询 HandlerMapping 找到请求对应的特定控制器，并将请求委托给该特定控制器；

(3) 特定控制器调用业务层组件完成业务逻辑；返回一个 ModelAndView 给 DispatcherServlet。ModelAndView 中包含视图逻辑名和模型数据；

(4) DispatcherServlet 使用 ViewResolver 完成逻辑视图到物理视图的解析；

(5) DispatcherServlet 将模型数据放入视图并发送视图到客户端。

13.1 Spring MVC 配置

配置 Spring MVC 有两种方式：使用 XML 配置 Spring MVC、使用 Java 配置 Spring MVC。

13.1.1 使用 XML 配置 Spring MVC

1. 配置 DispatcherServlet

在 Java Web 层部署描述文件 web.xml 中配置 Spring Web 核心控制器 DispatcherServlet 如下：

```
<?xml version="1.0" encoding="UTF-8"?>
<web-app …>
    …
    <servlet>
```

```
        <servlet-name>dispatcherServlet</servlet-name>
        <servlet-class>org.springframework.web.servlet.DispatcherServlet
</servlet-class>
        <load-on-startup>1</load-on-startup>
    </servlet>
    <servlet-mapping>
        <servlet-name>dispatcherServlet</servlet-name>
        <url-pattern>/</url-pattern>
    </servlet-mapping>
</web-app>
```

<servlet>标签用于声明 DispatcherServlet。

<servlet-mapping>指定 DispatcherServlet 处理的 URL 为/,即当前 Web 应用根路径下所有的请求都由 Spring 的 DispatcherServlet 处理(包括对静态资源的请求)。可以修改 DispatcherServlet 处理 URL 为.do 或.htm 等。

2. 指定 Spring Web 配置文件

Spring Web 配置文件描述了控制器、视图解析器等信息。

在启动 Spring Web 层容器时,DispatcherServlet 会加载 Spring Web 配置文件。DispatcherServlet 默认会加载/WEB-INF/下的 DispatcherServlet 的名字-servlet.xml。此处即加载 dispatcherServlet-servlet.xml 文件。

如果需要修改加载的 Spring Web 配置文件名字,可以使用配置 DispatcherServlet 时的子标签<init-param>和<param-value>指定。如下代码使用 namespace 参数名和 spring-mvc 参数值指定了 Spring Web 配置文件名字为 spring-mvc.xml。

```
<?xml version="1.0" encoding="UTF-8"?>
<web-app …
    <servlet>
        <servlet-name>dispatcherServlet</servlet-name>
        <servlet-class>org.springframework.web.servlet.DispatcherServlet
</servlet-class>
        <init-param>
            <param-name>namespace</param-name>
            <param-value>spring-mvc</param-value>
        </init-param>
        …
    </servlet>
        …
</web-app>
```

还可以使用如下 contextConfigLocation 配置加载的 Spring Web 配置文件的路径:

```
<?xml version="1.0" encoding="UTF-8"?>
<web-app …
```

```
<servlet>
    <servlet-name>dispatcherServlet</servlet-name>
    <servlet-class>org.springframework.web.servlet.DispatcherServlet
        </servlet-class>
    <init-param>
        <param-name>contextConfigLocation</param-name>
        <param-value>/WEB-INF/spring-mvc.xml</param-value>
    </init-param>
    <load-on-startup>1</load-on-startup>
</servlet>
  ...
</web-app>
```

3. 指定后端 Bean 配置文件

通常 Spring Web 应用中有一个由监听器 ContextLoadListener 创建的根应用上下文,该上下文负责加载 Web 应用后端(如 Service 层、Dao 层)的 Bean。在 web. xml 中配置监听器 ContextLoadListener 如下:

```
<?xml version="1.0" encoding="UTF-8"?>
<web-app>
  ...
  <context-param>
      <param-name>contextConfigLocation</param-name>
      <param-value>/WEB-INF/applicationContext.xml</param-value>
  </context-param>
  <listener>
      <listener-class>
          org.springframework.web.context.ContextLoaderListener
      </listener-class>
  </listener>
  ...
</web-app>
```

<listener>用于配置监听器 org. springframework. web. context. ContextLoaderListener。该监听在 Web 容器启动时会创建 Spring 根应用上下文,并默认加载/WEB-INF/applicationContext. xml 文件中描述的 Bean。

<context-param>配置上下文参数。<param-name>指定 Spring 根应用上下文常量 contextConfigLocation;<param-value>指定后端 Bean 配置文件 applicationContext. xml 的路径。

4. 配置 Spring Web 配置文件

描述了控制器、视图解析器等信息的 Spring Web 配置文件 spring-mvc. xml 配置如下:

```
<beans xmlns="http://www.springframework.org/schema/beans"
    xmlns:context="http://www.springframework.org/schema/context"
    xmlns:xsi="http://www.w3.org/2001/XMLSchema-instance"
    xsi:schemaLocation="
    http://www.springframework.org/schema/beans
    http://www.springframework.org/schema/beans/spring-beans-3.0.xsd
    http://www.springframework.org/schema/context
    http://www.springframework.org/schema/context/spring-context-3.0.xsd">
    <context:component-scan base-package="web" />
    <bean class=
        "org.springframework.web.servlet.view.InternalResourceViewResolver">
        <property name="prefix" value="/WEB-INF/views/"/>
        <property name="suffix" value=".jsp"/>
    </bean>
</beans>
```

xmlns:context 等加粗代码用于引入 Spring context 命名空间。

<context:component-scan base-package="web" />配置 Spring 自动扫描 web 包中使用@Component 注解的类，并为每个使用@Component 注解的类创建 Bean 和完成自动装配。

<bean class="org.springframework.web.servlet.view.InternalResourceViewResolver">配置 Spring 视图解析器为 InternalResourceViewResolver。InternalResourceViewResolver 视图解析器完成逻辑视图到 JSP 物理视图的解析工作。例如，某个处理器返回 String 的逻辑视图名 hello 时，InternalResourceViewResolver 视图解析器会在 hello 的前后分别加上前缀名称/WEB-INF/views/和后缀名称.jsp，形成/WEB-INF/views/hello.jsp 路径，完成逻辑视图到 JSP 物理视图的映射。

5. 配置后端 Bean 配置文件

Spring 应用上下文加载的后端 Bean 配置文件 applicationContext.xml 如下：

```
<?xml version="1.0" encoding="UTF-8"?>
<beans xmlns="http://www.springframework.org/schema/beans"
    xmlns:xsi="http://www.w3.org/2001/XMLSchema-instance"
    xmlns:context="http://www.springframework.org/schema/context"
    xmlns:aop="http://www.springframework.org/schema/aop"
    xsi:schemaLocation="http://www.springframework.org/schema/beans
    http://www.springframework.org/schema/beans/spring-beans-4.0.xsd
    http://www.springframework.org/schema/aop
    http://www.springframework.org/schema/aop/spring-aop-4.0.xsd
    http://www.springframework.org/schema/context
    http://www.springframework.org/schema/context/spring-context-4.0.xsd" >
    <context:component-scan base-package="service,dao" />
</beans>
```

<context: component-scan base-package = " service, dao" />指定 Spring 扫描 service 和 dao 包中带有@Component 注解的类，并为每个类创建 Bean 和完成自动装配。

13.1.2 使用 Java 配置 Spring MVC

1. 配置 DispatcherServlet

使用 Java 配置 DispatcherServlet 的 WebAppInit 类的关键代码如下：

```
package config;
...
public class WebAppInit extends AbstractAnnotationConfigDispatcherServletInitializer {
    @Override
    protected String[] getServletMappings() {
        return new String[] { "/" }; // 将 DispatcherServlet 映射到"/"
    }
}
```

WebAppInit 继承了 AbstractAnnotationConfigDispatcherServletInitializer，Java Web 容器会使用该类作为 Java Web 应用的配置类，即该类的作用等价于 web.xml。

getServletMappings()方法返回一个数组，该数组中的每个元素表示路径信息。对数组中每个路径的请求都交由 DispatcherServlet 处理。此处返回/表示该 Web 应用下所有的请求都由 DispatcherServlet 处理。

2. 指定 Spring Web 配置类

在 WebAppInit 类增加如下 getServletConfigClasses()方法：

```
package config;
...
public class WebAppInit extends AbstractAnnotationConfigDispatcherServletInitializer {
    // 指定 Spring Web 配置类
    @Override
    protected Class<?>[] getServletConfigClasses() {
        return new Class<?>[] { WebConfig.class };
    }
    @Override
    protected String[] getServletMappings() {
        ...
    }
}
```

getServletConfigClasses()方法返回的是 Spring Web 的配置类 WebConfig。配置类 WebConfig 描述了控制器、视图解析器等信息。DispatcherServlet 启动时会创建 Spring Web 应用上下文并加载配置类 WebConfig 中声明的 Bean。

3. 指定后端 Bean 配置类

在 WebAppInit 类增加如下 getRootConfigClasses()方法：

```
package config;
…
public class WebAppInit extends AbstractAnnotationConfigDispatcherServletInitializer {
    // 指定 Spring Web 以外的其他配置类，如 Service 层、Dao 层的配置类
    @Override
    protected Class<?>[] getRootConfigClasses() {
        return new Class<?>[] { RootConfig.class };
    }
    // 指定 Spring Web 配置类
    @Override
    protected Class<?>[] getServletConfigClasses() {
        return new Class<?>[] { WebConfig.class };
    }
    @Override
    protected String[] getServletMappings() {
        return new String[] { "/" }; // 将 DispatcherServlet 映射到"/"
    }
}
```

getRootConfigClasses()方法返回的是后端 Bean 的 Java 配置类 RootConfig。该配置类描述了 Spring 根应用上下文需要加载的 Web 应用后端（如 Service 层、Dao 层）的 Bean。

4. 配置 Spring Web 配置类

Spring Web 的配置类 WebConfig 的关键代码如下：

```
package config;
…
@Configuration
@EnableWebMvc // 启用 Spring MVC
@ComponentScan("web") // 启用组件扫描
public class WebConfig extends WebMvcConfigurerAdapter {
    @Bean
    public ViewResolver viewResolver() {
        InternalResourceViewResolver resolver =new InternalResourceViewResolver();
        resolver.setPrefix("/WEB-INF/views/");
        resolver.setSuffix(".jsp");
        resolver.setExposeContextBeansAsAttributes(true);
        return resolver;
    }
```

```
    // 配置静态资源的处理
    @Override
    public void configureDefaultServletHandling
        (DefaultServletHandlerConfigurer configurer) {
        configurer.enable();
    }
}
```

@Configuration 注解用于类前，指定该类是配置类。此处指定了 WebConfig 是配置类。

@EnableWebMvc 注解指定启用 Spring MVC。

@ComponentScan("web")用于启用组件扫描。Spring 会扫描 web 包中带有@Component 注解的类并创建这些类的 Bean 对象。

viewResolver()方法配置的 Bean 指定 Spring 视图解析器为 InternalResourceViewResolver。

configureDefaultServletHandling()方法配置对静态资源的处理由容器默认的 Servlet 处理，而不是 DispatcherServlet 处理。

5. 配置后端 Bean 配置类

Spring 根应用上下文加载的后端 Bean 配置类 RootConfig，其关键代码如下：

```
package config;
…
@Configuration
@ComponentScan(basePackages ={ "service", "dao" },
    excludeFilters ={
        @Filter(type =FilterType.ANNOTATION, value =EnableWebMvc.class) })
public class RootConfig {
}
```

@Configuration 注解指定该类为配置类。

@ComponentScan 注解用于开启组件扫描。其 basePackages 属性指定扫描包为 service 和 dao 包。excludeFilters 属性指定扫描排除，通过@Filter 指定对使用 EnableWebMvc 注解的配置类 WebConfig 中指定的扫描包 web 不会进行扫描。

13.2 编写控制器

13.2.1 第一个简单的控制器

编写一个简单控制器，实现在浏览器输入 http://localhost:8080/ch14-1-1/user/hello，Spring 调用控制器 UserController 的 hello()方法将物理视图 hello.jsp 发送到在浏览器显示。控制器 UserController 类的关键代码如下：

```
package web;
...
@Controller
@RequestMapping("/user")
public class UserController {
    @RequestMapping("/hello")
    public String hello() {
        return "user/hello";
    }
}
```

UserController 类前使用@Controller 注解表示该类是一个 Spring 控制器。Spring 自动扫描后会创建该类的 Bean。

@RequestMapping("/user")注解表示将该类的 Bean 映射到/user 路径。还可以使用其标准形式实现等价映射：@RequestMapping(value = "/user")。

hello()方法前的@RequestMapping("/hello")表示该方法映射到/hello。方法前的@RequestMapping 路径映射是相对于类前路径映射的。例如，上述 hello()方法映射路径是到/user/hello。

可以删除类前的@RequestMapping 注解，在 hello()方法上使用注解实现映射/user/hello 的关键代码如下：

```
package web;
...
@Controller
public class UserController {
    @RequestMapping("/user/hello")
    public String hello() {
        return "user/hello";
    }
}
```

hello()方法返回的逻辑视图名为 user/hello。根据配置类 WebConfig 中配置的 JSP 视图解析器，该逻辑视图被映射到/WEB-INF/views/user/hello. jsp。hello. jsp 的关键代码如下：

```
<body>
    hello spring!
</body>
```

13. 2. 2　处理请求参数

控制器处理客户端传递的请求参数也是一项常见操作。例如，在用户列表页面 listAll. jsp 的操作栏，单击“详细”将获得某个 id 的用户详细信息。

```
<td width="200">用户名</td><td width="200">操作</td>
<td><a href="${ctx}/user/get?id=${user.id}">详细</a></td>
```

控制器 UserController 的 get()方法需要处理对/user/get? id=${user.id}的请求并处理 id 参数。UserController 的关键代码如下：

```
...
@Controller
@RequestMapping("/user")
public class UserController {
    @Autowired
    private UserService userService;
    ...
    @RequestMapping("/get")
    public String get(@RequestParam("id") int id, Model model) {
        model.addAttribute("user", userService.getUserById(id));
        return "user/userInfo";
    }
}
```

在 UserController 的 get()方法中，@RequestParam("id") int id 表示请求参数 id 的值被 Spring 转型为 int id 来保存。

get()方法的第二个参数 Model 用于向前台页面传递数据，在 JSP 页面可以使用 JSP EL 获取。此处使用 model.addAttribute("user", userService.getUserById(id))表示将服务层获取的用户信息对象放置到 model 中。

显示用户详细信息的 userInfo.jsp 的关键代码如下：

```
...
<html>
<head><title>用户信息 userInfo.jsp</title></head>
<body>
用户信息<br/>
用户 ID=${user.id };用户名=${user.name};
</body>
</html>
```

userInfo.jsp 中使用 JSP EL ${user.id }和${user.name}显示用户 id 和用户名。

13.2.3 处理路径参数

从面向资源的角度，将查询参数/user/get? id=${user.id}变成如下路径参数更为合理：

```
<td><a href="${ctx}/user/getById/${user.id}">详细</a></td>
```

控制器 UserController 增加 getById()方法来处理上述路径参数，其关键代码如下：

```
...
@Controller
@RequestMapping("/user")
public class UserController {
    @Autowired
    private UserService userService;
    ...
    @RequestMapping(value ="/getById/{id}", method =RequestMethod.GET)
    public String getById(@PathVariable("id") int id, Model model) {
        model.addAttribute("user", userService.getUserById(id));
        return "user/userInfo";
    }
}
```

@RequestMapping 的 value = "/getById/{id}"使用占位符指定 id 作为路径的一部分。

getById 方法的第一个参数使用@PathVariable("id") int id 表示使用 int id 保存占位符 id 的值。

getById 方法的第二个参数 Model 用于向前台页面传递数据，在 JSP 页面可以使用 JSP EL 获取。userInfo.jsp 代码不变。

13.2.4 处理表单参数

控制器处理表单参数也是常见操作。下面以常见的用户注册为例说明控制器处理表单参数。在 listAll.jsp 增加 JSP 包含指令 include 以引入 JSP JSTL 标签库，从而在该页面使用<c:if>、<c:forEach>等标签。

```
<%@include file="../commons/jstl.jsp" %>
```

jstl.jsp 使用 taglib 指令代码如下：

```
<%@taglib prefix="c" uri="http://java.sun.com/jsp/jstl/core"%>
<%@taglib prefix="fmt" uri="http://java.sun.com/jsp/jstl/fmt"%>
<%@taglib prefix="fn" uri="http://java.sun.com/jsp/jstl/functions"%>
<c:set var="ctx" value="${pageContext.request.contextPath}"/>
```

这种将标签库文件放到一个文件中的方式便于代码维护和扩展。

listAll.jsp 页面新增一个注册超链接。

```
...
<a href="${ctx}/user/register">注册</a>    用户列表
...
```

用户单击注册超链接将调用 UserController 的 register()方法打开注册页面

registerForm.jsp。UserController 的 register()方法的关键代码如下：

```
@Controller
@RequestMapping("/user")
public class UserController {
    @Autowired
    private UserService userService;
    …
    // 打开注册页面
    @RequestMapping("/register")
    public String register() {
        return "user/registerForm";
    }
}
```

用户输入注册数据用户名的 registerForm.jsp 的关键代码如下：

```
…
<body>
注册
<form action="${ctx}/user/register" method="post">
  用户名:<input type="text" name="name"/><br/>
  <input type="submit" value="提交"/>
</form>
</body>
```

用户输入用户名后单击“提交”按钮，将注册数据交由 UserController 控制器的 doRegister 处理，其关键代码如下：

```
package web;
…
@Controller
@RequestMapping("/user")
public class UserController {
    @Autowired
    private UserService userService;
    // 打开注册页面
    @RequestMapping("/register")
    public String register() {
        return "user/registerForm";//
    }
    // 处理注册数据
    @RequestMapping(value="/register", method=RequestMethod.POST)
    public String doRegister(User user) {
        userService.add(user);
        return "redirect:/user/listAll";
```

```
    }
    @RequestMapping("/listAll")
    public String listAll(Model model) {
        model.addAttribute("allUsers", userService.listAll());
        return "user/listAll";
    }
}
```

redirect:/user/listAll 表示使用重定向跳转到用户列表页面。使用重定向可以防止用户刷新浏览器页面导致多次重复提交。注册成功后显示用户列表页面如图 13-1 所示。

用户id	用户名	操作
1	user-1	详细
2	user-2	详细
3	user-3	详细
4	user-4	详细

图 13-1　注册成功后显示用户列表页面

13.3　数据校验

为了保证用户输入数据的正确性和有效性，通常应用程序要对用户输入的数据进行校验。应用程序的数据校验分为客户端数据校验和服务器端数据校验。Web 应用程序中客户端数据校验使用 JavaScript 完成。为了避免用户绕过浏览器，使用 HTTP 工具直接向后端请求一些违法数据，服务端的数据校验也是必要的。本节讨论的数据校验属于服务器端的数据校验。

Java EE 6 包含的 JSR303 是 Java 的校验规范。JSR-349 是 JSR303 的升级版本，添加了一些新特性。JSR303/ JSR-349 通过在 Bean 上使用校验注解如@NotNull、@Max 等实现对 Bean 的验证，这些注解位于 javax.validation.constraints 包。JSR303/ JSR-349 校验注解说明见表 13-1。

表 13-1　JSR303/ JSR-349 校验注解说明

注　解	说　明
@Null	被注解的元素必须为 null
@NotNull	被注解的元素必须不为 null
@AssertTrue	被注解的元素必须为 true
@AssertFalse	被注解的元素必须为 false

续表

注　　解	说　　明
@Min(value)	被注解的元素必须是数字且大于或等于指定值 value
@Max(value)	被注解的元素必须是数字且小于或等于指定值 value
@DecimalMin(value)	被注解的元素必须是数字且大于或等于指定值 value
@DecimalMax(value)	被注解的元素必须是数字且小于或等于指定值 value
@Size(max, min)	被注解的元素的大小必须在指定的 min 和 max 之间
@Digits (integer, fraction)	被注解的元素必须是数字且整数部分、小数部分满足指定值要求
@Past	被注解的元素必须是过去的日期
@Future	被注解的元素必须是将来的日期
@Pattern(value)	被注解的元素必须符合指定的正则表达式

Hibernate Validation 验证框架是 Java 校验规范的一个实现，并增加了一些其他校验注解如@Email、@Length、@Range 等。Spring 对 Hibernate Validation 进行了封装，在 Spring MVC 模块中添加了自动校验，并将校验信息封装进了特定的类中，便于 Web 应用程序的快速开发。Hibernate 校验注解说明见表 13-2。

表 13-2　Hibernate 校验注解说明

注　　解	说　　明
@Email	被注解的元素必须是电子邮箱地址
@Length	被注解的字符串的大小必须在指定的范围内
@NotEmpty	被注解的字符串的必须非空
@Range	被注解的元素必须在合适的范围内

下面介绍使用 Spring 完成数据校验：注册页面中用户名不能为空、用户名字符数为 3～20(包括 3 和 20)。本案例见项目 ch13-1-2，需要复制 javax. validation-1. 0. 0. GA. jar 和 hibernate-validator-4. 3. 1. final-1. 0. 0. jar 到项目 lib 下。用户注册时填写的输入数据不满足要求时页面显示如图 13-2 所示。

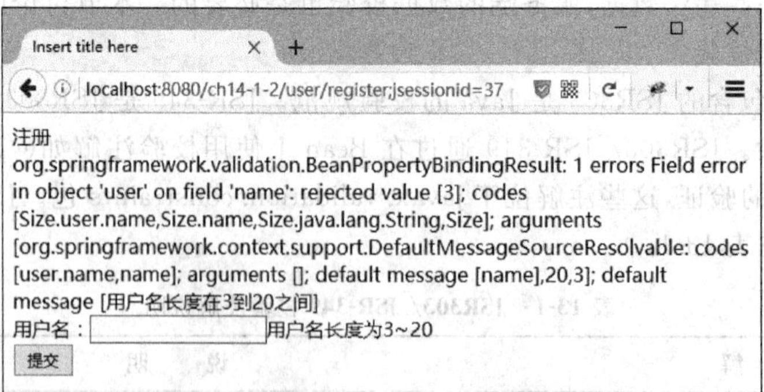

图 13-2　用户注册时填写的输入数据不满足要求时页面显示

1. 在类 User 的 name 属性上使用@NotNull 注解和@Size 注解

```
package entity;
import javax.validation.constraints.NotNull;
import javax.validation.constraints.Size;
public class User {
    private int id;
    @NotNull(message ="用户名不能为空")
    @Size(min =3, max =20, message ="用户名长度为 3~20")
    private String name;
    public User() {
    }
    …
}
```

2. 在控制器 doRegister()方法上添加校验注解@Valid 启用校验功能

```
package web;
…
@Controller
@RequestMapping("/user")
public class UserController {
    @Autowired
    private UserService userService;
    …
    // 打开注册页面
    @RequestMapping("/register")
    public String register() {
        return "user/registerForm";//
    }
    // 处理注册数据
    @RequestMapping(value ="/register", method=RequestMethod.POST)
    // public String doRegister(@Valid User user,
    // BindingResult result, Model model) {
    public String doRegister(@Valid User user, Errors errors, Model model) {
        if (errors.hasErrors()) {
            model.addAttribute("errors", errors);
            return "/user/registerForm";
        }
        userService.add(user);
        return "redirect:/user/listAll";
    }
    …
}
```

doRegister 方法的 User user 参数前面添加了@Valid 注解，表示需要按照校验规则对 user 对象进行校验。如果有校验出现错误，则使用 model 保存错误并返回到注册页面 registerForm，让用户修正错误，然后重新提交。如果没有错误，则调用 userService. add (user)保存注册数据并重定向到用户列表页面。

使用显示错误信息的注册页面 registerForm. jsp 的关键代码如下：

```
<%@page language="java" contentType="text/html; charset=UTF-8"
    pageEncoding="UTF-8"%>
<%@taglib prefix="sf" uri="http://www.springframework.org/tags/form"%>
<html>
…
<body>
注册
<%--<form action="${ctx}/user/register" method="post">--%>
<sf:form method="post" modelAttribute="user">
  <span>${errors}</span>
  用户名:<input type="text" name="name"/>
     <span><sf:errors path="name"/></span><br/>
  <input type="submit" value="提交"/>
</sf:form>
</body>
</html>
```

注册页面 registerForm. jsp 使用 taglib 指定导入 Spring 的标签库 form。该页面的 form 表单中使用 modelAttribute="user"绑定了模型对象 user。<span>${errors}</span>显示所有字段的校验出错信息。如果需要显示某个字段的出错信息，可以在输入域的面使用 sf:errors path="name"/>输出 name 校验的出错信息。

Spring MVC 支持错误消息的国际化，限于篇幅在此不予介绍。

13.4 视图解析

Spring MVC 处理视图最重要的两个接口是 ViewResolver 和 View。ViewResolver 将控制器返回的逻辑视图解析为物理视图；View 将真正的视图呈现给用户。

Spring 常见视图解析器见表 13-3。表中的每一种视图解析器都对应 Java Web 应用中特定的某种视图技术。实际应用开发中通常只会用到少数几种视图解析器：InternalResourceViewResolver 一般会用于 JSP；TilesViewResolver 用于 Apache Tiles 视图；FreeMarkerViewResolver 和 VelocityViewResolver 分别对应 FreeMarker 和 Velocity 模板视图。

表 13-3　Spring 常见视图解析器

视图解析器	说　明
BeanNameViewResolver	将视图解析为 Spring 应用上下文中的 Bean,其中 Bean 的 ID 与视图的名字相同
ContentNegotiatingViewResolver	通过考虑客户端需要的内容类型来解析视图,委托给另外一个能够产生对应内容类型的视图解析器
FreeMarkerViewResolver	将视图解析为 FreeMarker 模板
InternalResourceViewResolver	将视图解析为 Web 应用的内部资源(一般为 JSP)
JasperReportsViewResolver	将视图解析为 JasperReports 定义
ResourceBundleViewResolver	将视图解析为资源 bundle(一般为属性文件)
TilesViewResolver	将视图解析为 Apache Tile 定义,其中 tile ID 与视图名称相同。有两个不同的 TilesViewResolver 实现,分别对应于 Tiles 2.0 和 Tiles 3.0
UrlBasedViewResolver	直接根据视图的名称解析视图,视图的名称会匹配一个物理视图的定义
VelocityLayoutViewResolver	将视图解析为 Velocity 布局,从不同的 Velocity 模板中组合页面
VelocityViewResolver	将视图解析为 Velocity 模板
XmlViewResolver	将视图解析为特定 XML 文件中的 Bean 定义。类似于 BeanNameViewResolver
XsltViewResolver	将视图解析为 XSLT 转换后的结果

13.4.1　JSP 视图

Spring 提供了两种支持 JSP 视图的方式。

- InternalResourceViewResolver。将逻辑视图解析为 JSP 文件。还用于将逻辑视图解析为 JstlView 形式的 JSP 文件,这样可以在 JSP 页面中使用 JSTL。
- Spring 标签库。Spring 自带两个 JSP 标签库:一个用于表单到模型的绑定,另一个作为通用的工具类。

配置 JSP 视图解析器包括两种方式。

1. 使用 Java 配置 JSP 视图解析器

使用 Java 配置 JSP 视图解析器,配置类 WebConfig 的关键代码如下:

```
package config;
...
@Configuration
@EnableWebMvc // 启用 Spring MVC
@ComponentScan("web") //启用组件扫描
```

```
public class WebConfig extends WebMvcConfigurerAdapter {
    @Bean
    public ViewResolver viewResolver() {
        InternalResourceViewResolver resolver =new InternalResourceViewResolver();
        resolver.setPrefix("/WEB-INF/views/");
        resolver.setSuffix(".jsp");
        resolver.setViewClass(org.springframework.web.servlet.view.JstlView.
class);
        resolver.setExposeContextBeansAsAttributes(true);
        return resolver;
    }
    //配置静态资源的处理
    @Override
    public void configureDefaultServletHandling
        (DefaultServletHandlerConfigurer configurer) {
        configurer.enable();
    }
}
```

在配置类 WebConfig 中，resolver. setPrefix 和 setSuffix 分别为逻辑视图添加前缀和后缀。例如，控制器返回的逻辑视图为 home，InternalResourceViewResolver 会在 home 前后分别增加前缀/WEB-INF/views/和后缀. jsp，这样产生物理视图即 JSP 文件位置为/WEB-INF/views/home. jsp。setViewClass 方法配置将逻辑视图解析为 JstlView 形式的 JSP 文件。

2. 使用 XML 配置 JSP 视图解析器

XML 配置文件的关键代码如下：

```
<?xml version="1.0" encoding="UTF-8"?>
<beans …>
   <annotation-driven />
   <context:component-scan base-package="web" />
   <beanclass=
      "org.springframework.web.servlet.view.InternalResourceViewResolver">
      <property name="prefix" value="/WEB-INF/views/" />
      <property name="suffix" value=".jsp" />
      <property name="viewClass"
      value="org.springframework.web.servlet.view.JstlView"/>
   </bean>
</beans>
```

13.4.2 Tile 视图

实际的 Web 项目开发中很多页面具有相同的外观。例如，每个页面都需要有

header、menu、body、footer 等，如图 13-3 所示。

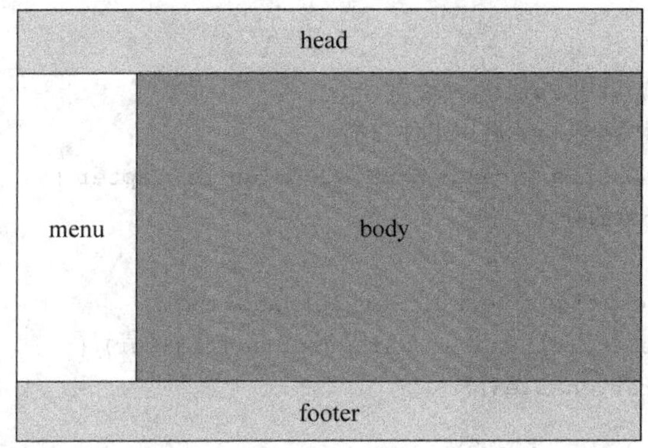

图 13-3 具有 header、menu、body、footer 相同外观的页面

Apache Tiles 是一个 Java EE 应用的页面布局框架。Tiles 框架提供了一种模板机制，可以为某一类页面定义一个通用的模板，该模板定义了页面的整体布局。布局由可以复用的多个块组成，每个页面可以有选择性的重新定义块而达到组件的复用。

Spring 提供了视图解析器支持将逻辑视图转换为 Tiles 视图。下面使用 Spring MVC 整合 Apache Tile 视图解析。该应用程序具有相同的 header、menu、footer，右方的内容区域不同。显示效果如图 13-4 所示。

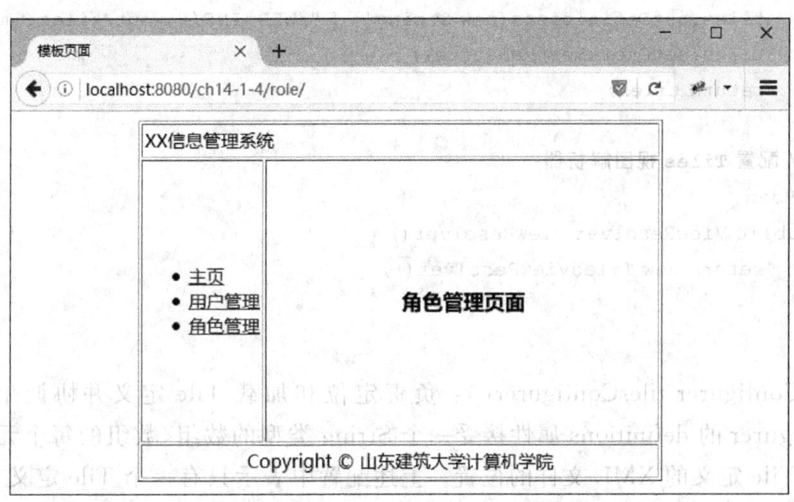

图 13-4 具有相同外观的页面显示效果

1. 配置 Tiles 视图解析器

配置 Tiles 视图解析器可以使用 Java 配置，也可以使用 XML 配置。在项目中选择一种配置方式即可。

1）使用 Java 配置类配置 Tiles 视图解析器

在配置类 WebConfig 中配置 Tiles 视图解析器的关键代码如下：

```
package config;
...
@Configuration
@EnableWebMvc // 启用 Spring MVC
@ComponentScan("web") // 启用组件扫描
public class WebConfig extends WebMvcConfigurerAdapter {
    // 配置静态资源的处理
    @Override
    public void configureDefaultServletHandling
        (DefaultServletHandlerConfigurer configurer) {
        configurer.enable();
    }
    @Override
    public void addResourceHandlers(ResourceHandlerRegistry registry) {
        // TODO Auto-generated method stub
        super.addResourceHandlers(registry);
    }
    // 配置 TilesConfigurer
    @Bean
    public TilesConfigurer tilesConfigurer() {
        TilesConfigurer tiles =new TilesConfigurer();
        tiles.setDefinitions(new String[] { "/WEB-INF/layout/tiles.xml" });
        tiles.setCheckRefresh(true);
        return tiles;
    }
    // 配置 Tiles 视图解析器
    @Bean
    public ViewResolver viewResolver() {
        return new TilesViewResolver();
    }
}
```

TilesConfigurer tilesConfigurer()：负责定位和加载 Tile 定义并协调生成 Tiles。TilesConfigurer 的 definitions 属性接受一个 String 类型的数组，数组的每个元素都指定一个表示 Tile 定义的 XML 文件的位置。上述配置中表示只有一个 Tile 定义文件 tiles.xml，该文件位于 WEB-INFlayout/目录下。

可以在此指定存在多个 Tile 定义文件。例如，如下代码指定 TilesConfigurer 加载 WEB-INF 目录下所有的名为 tiles.xml 的文件。

```
tiles.setDefinitions(new String[] {"/WEB-INF/**/tiles.xml" });
```

ViewResolver viewResolver()：配置 Tiles 视图解析器将逻辑视图解析为 Tiles 视图。

2）使用 XML 配置 Tiles 视图解析器

在 WEB-INF/spring-mvc. xml 中使用 XML 配置 Tiles 视图解析器代码如下：

```
<!--配置 TilesConfigurer -->
<bean id="tilesConfigurer"
   class="org.springframework.web.servlet.view.tiles3.TilesConfigurer">
   <property name="definitions">
      <list>
         <value>/WEB-INF/layout/tiles.xml</value>
      </list>
   </property>
</bean>
<!-配置 Tiles 视图解析器 -->
<bean id="viewResolver"
   class="org.springframework.web.servlet.view.tiles3.TilesViewResolver" />
```

2. 配置 Tiles 文件

Tiles 配置文件 tiles. xml 配置如下：

```
<?xml version="1.0" encoding="ISO-8859-1" ?>
<!DOCTYPE tiles-definitions PUBLIC
        "-//Apache Software Foundation//DTD Tiles Configuration 3.0//EN"
        "http://tiles.apache.org/dtds/tiles-config_3_0.dtd">
<tiles-definitions>
   <definition name="base" template="/WEB-INF/layout/template.jsp">
      <put-attribute name="header" value="/WEB-INF/layout/header.jsp" />
      <put-attribute name="left" value="/WEB-INF/layout/left.jsp" />
      <put-attribute name="footer" value="/WEB-INF/layout/footer.jsp" />
   </definition>
   <definition name="home" extends="base">
      <put-attribute name="body" value="/WEB-INF/views/home.jsp" />
   </definition>
   <definition name="user" extends="base">
      <put-attribute name="body" value="/WEB-INF/views/user/user.jsp" />
   </definition>
   <definition name="role" extends="base">
      <put-attribute name="body" value="/WEB-INF/views/role/role.jsp" />
   </definition>
</tiles-definitions>
```

每个 Tile 定义中使用一个<definition>元素，该元素会有一个或多个<put-attribute>元素。

每个<definition>元素都定义了一个 Tile，它最终引用的是一个 JSP 模板。在名为 base 的 Tile 中，模板引用的是 WEB-INFlayout/ template. jsp。某个 Tile 可能还会引用其他的 JSP 模板。

3. 编写相关页面

(1) 模板页面 template.jsp 关键代码如下：

```
<%@taglib uri="http://www.springframework.org/tags" prefix="s"%>
<%@taglib uri="http://tiles.apache.org/tags-tiles" prefix="tiles"%>
…
<html><head>
<title>模板页面</title></head>
<body>
  <table border="1" cellpadding="2" cellspacing="2" align="center">
     <tr>
       <td height="30" colspan="2"><tiles:insertAttribute name="header" />
       </td>
     </tr>
     <tr>
       <td height="250"><tiles:insertAttribute name="left" /></td>
       <td width="350"><tiles:insertAttribute name="body" /></td>
     </tr>
     <tr>
       <td height="30" colspan="2"><tiles:insertAttribute name="footer" />
       </td>
     </tr>
  </table>
</body>
</html>
```

(2) 头部文件 header.jsp 的关键代码如下：

```
<%@page session="false" pageEncoding="UTF-8"
   contentType="text/html; charset=utf-8"%>
<%@taglib uri="http://www.springframework.org/tags" prefix="s" %>
XX 信息管理系统
```

(3) 左侧导航菜单页面 left.jsp 的关键代码如下：

```
<%@page session="false" pageEncoding="UTF-8"
     contentType="text/html; charset=utf-8"%>
<nav>
  <ul id="menu">
     <li><a href="${pageContext.request.contextPath}/home">主页</a></li>
     <li><a href="${pageContext.request.contextPath}/user/">用户管理</a></li>
     <li><a href="${pageContext.request.contextPath}/role/">角色管理</a></li>
  </ul>
</nav>
```

（4）页脚页面 footer.jsp 的关键代码如下：

```
<%@page session="false" pageEncoding="UTF-8"
    contentType="text/html; charset=utf-8"%>
<div align="center">Copyright &copy; 山东建筑大学计算机学院</div>
```

（5）主页 home.jsp 关键代码如下：

```
<%@page session="false" pageEncoding="UTF-8" contentType="text/html;
  charset=utf-8"%>
<%@taglib uri="http://java.sun.com/jsp/jstl/core" prefix="c" %>
<%@page session="false" %>
<div align="center"><h3>主页内容</h3></div>
```

限于篇幅，其他页面 user.jsp、listAll.jsp、role.jsp 在此没有列出代码，请查看相关项目文件。

4. 编写控制器

HomeController 的关键代码如下：

```
package web;
…
@Controller
@RequestMapping("/")
public class HomeController {
   @RequestMapping(value = { "/", "/home" }, method =RequestMethod.GET)
   public String home(Model model) {
      return "home";
   }
}
```

13.4.3 返回 Json

Spring 4.0 MVC 的 MappingJackson2JsonView 使用 Jackson 框架的 ObjectMapper 将模型数据转换为 JSON 格式输出。在 spring-mvc.xml 中需要配置 MappingJackson2JsonView。

使用 XML 配置如下：

```
…
<bean id="jsonConverter"class=
  "org.springframework.http.converter.json.MappingJackson2HttpMessageConverter">
</bean>
<beanclass=
  "org.springframework.web.servlet.mvc.annotation.AnnotationMethodHandlerAdapter">
```

```
    <property name="messageConverters">
        <list>
            <ref bean="stringConverter" />
            <ref bean="jsonConverter" />
        </list>
    </property>
</bean>
...
```

在控制器中添加一个返回 userListJson 逻辑视图名的方法 userListInJson(Model model),其关键代码如下：

```
@RequestMapping(path="userList")
    public String userListInJson(Model model) {
    List<User>userList =new ArrayList<User>();
    ...
    model.addAttribute("userList", userList);
    return userListJson;
}
```

13.5 文件上传

DispatcherServlet 没有实现任何解析 multipart 请求数据的功能,而是将该任务委托给了 Spring 中 MultipartResolver 策略接口的实现,通过这个实现类来解析 multipart 请求中的内容。因此在 Spring 中实现文件上传需要配置 MultipartResolver。Spring 3.1 以上版本内置了两个 MultipartResolver 的实现。

- CommonsMultipartResolver：使用 Jakarta Commons FileUpload 解析 multipart 请求。
- StandardServletMultipartResolver：使用 Servlet 3.0 对 multipart 请求的支持。

通常优先选择配置 StandardServletMultipartResolver,因为其不需要依赖任何其他的项目。如果应用需要将部署到 Servlet 3.0 之前的容器或没有使用 Spring 3.1 以上更高版本,那么可能会需要 CommonsMultipartResolver。

下面介绍如何在 Spring MVC 中使用 Servlet 3.0 处理文件上传。

1. 配置 StandardServletMultipartResolver

在 Spring Web 配置类 WebConfig 中配置 StandardServletMultipartResolver 的 Bean 如下：

```
package config;
...
@Configuration
```

```
@EnableWebMvc // 启用 Spring MVC
@ComponentScan("web") // 启用组件扫描
public class WebConfig extends WebMvcConfigurerAdapter {
    @Bean
    public ViewResolver viewResolver() {
        …
    }
    // 配置静态资源的处理
    @Override
    public void configureDefaultServletHandling(
        DefaultServletHandlerConfigurer configurer) {
        configurer.enable();
    }
    // 配置 StandardServletMultipartResolver 是实现文件上传的前提
    @Bean
    public MultipartResolver multipartResolver() {
        return new StandardServletMultipartResolver();
    }
}
```

2. 配置上传文件配置信息

配置上传文件配置信息包括配置默认的上传文件路径、上传文件大小等。

如果配置 DispatcherServlet 的初始化类继承了 AbstractAnnotationConfigDispatcherServletInitializer 或 AbstractDispatcherServletInitializer，可以通过重载 customizeRegistration() 方法来配置默认的上传文件路径、上传文件大小等配置信息。Java 配置类 WebAppInit 实现了 AbstractAnnotationConfigDispatcherServletInitializer，配置上传文件不超过 2MB，整个请求数据不超过 4MB，配置代码如下：

```
public class WebAppInit extends AbstractAnnotationConfigDispatcherServletInitializer {
    …
    @Override
    protected String[] getServletMappings() {
        return new String[] { "/" }; // 将 DispatcherServlet 映射到"/"
    }
    @Override
    protected void customizeRegistration(Dynamic registration) {
        registration.setMultipartConfig(
            new MultipartConfigElement("/tmp/spittr/uploads", 2097152, 4194304, 0));
    }
}
```

使用 XML 等价的上传文件配置信息如下：

```
<servlet>
```

```
    <servlet-name>SpringDispatcher</servlet-name>
    <servlet-class>org.springframework.web.servlet.DispatcherServlet</servlet
-class>
    <multipart-config>
        <location>/tmp</location>
        <max-file-size>2097152</max-file-size><!--最大上次文件为 2MB-->
        <max-request-size>4194304</max-request-size><!--最大请求数据为 4MB-->
        <file-size-threshold>0</file-size-threshold>
    </multipart-config>
</servlet>
```

3. 编写上传文件页面

用户注册页面 registerForm.jsp 的关键代码如下：

```
<body>
注册
<%--<form action="${ctx}/user/register" method="post">--%>
<form action="${ctx}/user/doRegister" method="post" enctype="multipart/form-
data">
    <span>${errors}</span>
    用户名:<input type="text" name="name"/><br/>
    上传文件:<input type="file" name="uploadfile"/><br/>
    <input type="submit" value="提交"/>
</form>
</body>
```

<form>标签需要指定 enctype="multipart/form-data"。

4. 编写控制器实现文件上传

控制器 UserController 的关键代码如下：

```
package web;
…
@Controller
@RequestMapping("/user")
public class UserController {
    @Autowired
    private UserService userService;
    …
    // 打开注册页面
    @RequestMapping(value ="/register", method =RequestMethod.GET)
    public String register() {
        return "user/registerForm";//
    }
```

```
    // 使用 Spring 的 MultipartFile 处理上传文件
    @RequestMapping(value ="/doRegister", method =RequestMethod.POST)
    // public String doRegister(@Valid @ModelAttribute("user") User user,
    // BindingResult result, Model model) {
    public String doRegister(@RequestPart("uploadfile") MultipartFile uploadfile,
        @Valid User user, Errors errors,Model model, HttpServletRequest request) {
        if (!uploadfile.isEmpty()) {
            // 获得物理路径 webapp 所在路径
            String pathRoot =request.getServletContext().getRealPath("/uploads");
            String fileOriginalName =uploadfile.getOriginalFilename();
            fileOriginalName =fileOriginalName.substring(
                fileOriginalName.lastIndexOf("\\") +1);
            try {
                uploadfile.transferTo(new File(pathRoot +"\\" +fileOriginalName));
            } catch (IllegalStateException e) {
                // TODO Auto-generated catch block
                e.printStackTrace();
            } catch (IOException e) {
                // TODO Auto-generated catch block
                e.printStackTrace();
            }
        }
        if (errors.hasErrors()) {
            model.addAttribute("errors", errors);
            return "/user/registerForm";
        }
        userService.add(user);
        return "redirect:/user/listAll";
    }
    …
}
```

小　　结

本章首先介绍了 Spring MVC 的处理请求过程,Spring MVC 配置方式,如何编写控制器处理各种请求和使用数据校验。Spring 视图解析主要介绍了 JSP 视图解析器和 Tile 视图解析器以及返回 Json 数据。最后介绍了 Spring 文件上传时的配置,并编写了一个用户注册时上传文件的案例。

思考与习题

1. 写出 Spring 的核心控制器的全名(包名.类名)。
2. 写出 Spring MVC 处理请求的过程。
3. 配置 Spring MVC 的两种方式分别是什么?
4. 写出使用 XML 配置 Spring MVC 的主要内容。
5. 写出使用 Java 配置 Spring MVC 的主要内容。
6. 举例说明@RequestMapping 注解的用法。
7. 举例说明@RequestParam 注解的用法。
8. 使用 Spring MVC 编写用户登录程序。
9. 应用程序的数据校验分几种?
10. Hibernate Validation 验证框架与 Java 校验规范的关系是什么?
11. @Valid 注解的作用是什么?
12. 举例说明@NotNull 和@Size 注解的作用。
13. Spring 提供的两种支持 JSP 视图的方式分别是什么?
14. 写出使用 Java 配置 JSP 视图解析器时配置类 WebConfig 关键代码。
15. 写出使用 XML 配置 JSP 视图解析器时配置文件的关键代码。
16. 写出使用 Java 配置类配置 Tiles 视图解析器时配置类 WebConfig 的关键代码。
17. 写出使用 XML 配置 Tiles 视图解析器时的 spring-mvc.xml 的关键代码。
18. Spring 中实现文件上传需要配置什么?
19. Spring 3.1 以上版本内置的两个 MultipartResolver 实现分别是什么?
20. 写出使用 Java 和 XML 配置文件上传的关键代码。